D1715391

Springer Theses

Recognizing Outstanding Ph.D. Research

Aims and Scope

The series "Springer Theses" brings together a selection of the very best Ph.D. theses from around the world and across the physical sciences. Nominated and endorsed by two recognized specialists, each published volume has been selected for its scientific excellence and the high impact of its contents for the pertinent field of research. For greater accessibility to non-specialists, the published versions include an extended introduction, as well as a foreword by the student's supervisor explaining the special relevance of the work for the field. As a whole, the series will provide a valuable resource both for newcomers to the research fields described, and for other scientists seeking detailed background information on special questions. Finally, it provides an accredited documentation of the valuable contributions made by today's younger generation of scientists.

Theses are accepted into the series by invited nomination only and must fulfill all of the following criteria

- They must be written in good English.
- The topic should fall within the confines of Chemistry, Physics, Earth Sciences, Engineering and related interdisciplinary fields such as Materials, Nanoscience, Chemical Engineering, Complex Systems and Biophysics.
- The work reported in the thesis must represent a significant scientific advance.
- If the thesis includes previously published material, permission to reproduce this must be gained from the respective copyright holder.
- They must have been examined and passed during the 12 months prior to nomination.
- Each thesis should include a foreword by the supervisor outlining the significance of its content.
- The theses should have a clearly defined structure including an introduction accessible to scientists not expert in that particular field.

More information about this series at http://www.springer.com/series/8790

Linfei Li

Fabrication and Physical Properties of Novel Two-dimensional Crystal Materials Beyond Graphene: Germanene, Hafnene and $PtSe_2$

Doctoral Thesis accepted by
Institute of Physics, Chinese
Academy of Sciences, Beijing, China

Author
Dr. Linfei Li
Institute of Physics
Chinese Academy of Sciences
Beijing, China

Supervisor
Prof. Hong-jun Gao
Beijing, China

ISSN 2190-5053 ISSN 2190-5061 (electronic)
Springer Theses
ISBN 978-981-15-1962-8 ISBN 978-981-15-1963-5 (eBook)
https://doi.org/10.1007/978-981-15-1963-5

© Springer Nature Singapore Pte Ltd. 2020
This work is subject to copyright. All rights are reserved by the Publisher, whether the whole or part of the material is concerned, specifically the rights of translation, reprinting, reuse of illustrations, recitation, broadcasting, reproduction on microfilms or in any other physical way, and transmission or information storage and retrieval, electronic adaptation, computer software, or by similar or dissimilar methodology now known or hereafter developed.
The use of general descriptive names, registered names, trademarks, service marks, etc. in this publication does not imply, even in the absence of a specific statement, that such names are exempt from the relevant protective laws and regulations and therefore free for general use.
The publisher, the authors and the editors are safe to assume that the advice and information in this book are believed to be true and accurate at the date of publication. Neither the publisher nor the authors or the editors give a warranty, expressed or implied, with respect to the material contained herein or for any errors or omissions that may have been made. The publisher remains neutral with regard to jurisdictional claims in published maps and institutional affiliations.

This Springer imprint is published by the registered company Springer Nature Singapore Pte Ltd.
The registered company address is: 152 Beach Road, #21-01/04 Gateway East, Singapore 189721, Singapore

To my family,
for their endless love
and sustained support!

Supervisor's Foreword

The world is always progressing along with the development of new materials. Some of the significant materials have even been used to name the ages in ancient times, such as stone, bronze, and iron. In recent times, the discoveries of new materials usually bring about considerable influences on fundamental science. Today, it has been realized that the change in size and dimensionality of a known material, instead of chemical compositions, can also result in new material behaviors with exotic properties. Graphene, the first two-dimensional (2D) atomic crystal, is one of the most famous examples in this sense. The significance of graphene is not only its exceptional properties and enormous application potentials but more importantly, a crucial contribution to the establishment and advancement of a new research field—2D materials. Graphene's success and the methodology originally developed in graphene studies aroused intense interest in exploring other ultra-thin 2D materials, which has recently been at the leading edge of material science. This is documented by an exponential increase in publications in graphene-analogous material fields in the last dozen years. It is in this background, which we could call the post-graphene era, that the studies presented in this thesis have been performed.

Dr. Linfei Li, from my group, is the major contributor to the research. He and his collaborators devoted themselves to the investigation of novel 2D atomic crystals beyond graphene, involving preparation methods, structural characterizations, electronic properties, and potential applications. The state-of-the-art ultra-high vacuum molecular-beam-epitaxial scanning tunneling microscope (UHV-MBE-STM) systems combined with other advanced surface science techniques were employed, facilitating these experimental studies at a high level. The published original papers are of fairly high academic standard and collected in this thesis in a systematical way.

The present book approaches the topic in Chaps. 2–4 following the Introduction described in Chap. 1, which summarizes the most recent advancement in the field of 2D graphene-like materials, focusing on the introduction of growth methods, atomic and electronic structures, and application-related developments of several 2D crystals and their heterostructures. Chapter 2 reports studies on germanene, a germanium-based counterpart of graphene. I would like to stress that this is the first reported experimental work addressing the synthesis and structural properties of germanene sheets, which has received considerable attention (hundreds of citations) and initiates the corresponding experimental studies afterward. In Chap. 3, the authors challenge the conventional wisdom that honeycomb-like structures can exist only in elements that bear high chemical similarity to carbon, such as silicon (silicene) and germanium (germanene). They reveal that the honeycomb structure of transition metal (TM) elements can also be fabricated with totally uncharted physics and chemistry in particular for Hf on Ir(111) substrate. This unique TM honeycomb lattice, providing a new playground for investigating novel quantum phenomena and electronic behaviors, has sparked significant interest and been highlighted by Nature Nanotechnology and Nature China upon being reported. Chapter 4 deals with the pioneering experimental studies of single-layer $PtSe_2$, a new member of transition metal dichalcogenides (TMDs), which are a huge family of layered graphene-like materials. A simple approach through epitaxial growth on active metal substrates has been developed for high-quality monolayer $PtSe_2$ films. The atomic and band structures are demonstrated experimentally for the first time by a variety of high-resolution characterizations and first-principle theoretical calculations. The photocatalytic performance and circular polarization calculations underline the application potentials of monolayer $PtSe_2$ in photocatalysis and valleytronics. This work attracts more attention to other TMDs materials rather than the highly studied ones, such as MX_2 (M = Mo, W, X = S, Se, Te), and thus expands the scope of 2D atomic crystals. To the end, the important conclusions and instructive outlooks are given in Chap. 5.

In summary, this book involves the forefront of physics and material science and provides the reader with a systematical introduction to graphene-analogous 2D materials, one of the hottest topics in the last decade. Three novel atomic crystals, i.e., germanene, hafnene, and monolayer $PtSe_2$, are fabricated and experimentally studied for the first time. The reader will be led from the material preparations to the structural characterizations and from the study of properties to the exploration of applications. These original work could shed light on some significant issues in the field of 2D materials beyond graphene. I wish the book success in this direction.

Beijing, China
December 2019

Prof. Hong-jun Gao

Parts of this thesis have been published in the following journal articles:

The full articles were reprinted (adapted) with permission from ref. 1, © 2015 ACS; ref. 2, © 2014 Wiley; ref. 3, © 2013 ACS.

1. Yeliang Wang*, **Linfei Li*** (co-first author), Wei Yao, Shiru Song, J. T. Sun, Jinbo Pan, Shixuan Du, and Hong-Jun Gao# *et al.*, "Monolayer $PtSe_2$, a New Semiconducting Transition-Metal-Dichalcogenide, Epitaxially Grown by Direct Selenization of Pt", *Nano Letters 15, 4013 (2015)*. **[ESI highly cited paper]**
2. **Linfei Li***, Shuang-zan Lu*, Jinbo Pan, Zhihui Qin, Yu-qi Wang, Yeliang Wang, Geng-yu Cao, Shixuan Du, and Hong-Jun Gao#, "Buckled Germanene Formation on Pt(111)", *Advanced Materials 26, 4820 (2014)*. **[ESI highly cited paper]**
3. **Linfei Li**, Yeliang Wang, Shengyi Xie, Xianbin Li, Yu-Qi Wang, Rongting Wu, Hongbo Sun, Shengbai Zhang, and Hong-Jun Gao#, "Two-Dimensional Transition Metal Honeycomb Realized: Hf on Ir(111)", *Nano Letters 13, 4671 (2013)*. **[Highlighted by Nature Nanotechnology and Nature China]**

Acknowledgements

Time flies like an arrow. Six years have passed in a flash, and my Ph.D. career is coming to an end. In these exciting years, I met too many people who moved me, leaving behind lots of good memories. It was their help and care that promoted my personal growth and career development. I'm so lucky and happy to meet them, whom I'm very grateful to.

First of all, I would like to express the depth of my gratitude to my supervisor, Prof. Hong-jun Gao, for his careful guidance and instructive advice during my Ph.D. research. I am deeply grateful for his help in the completion of this thesis.

Second, grateful acknowledgment is made to my vice supervisor, Prof. Yeliang Wang, who taught me experimental techniques and writing skills, and shared his professional knowledge and useful suggestions. I feel lucky to have worked with him for six years.

Also, I would like to thank Prof. Shixuan Du, who offered me considerable help and valuable instructions in data discussions and theoretical calculations. I also owe a special debt of gratitude to all the other professors, staffs, and students in our N04 group. Thank you all for your kind help and support.

Besides, I am deeply indebted to all the collaborators from other groups. Thanks for their helpfulness and excellent expertise, which significantly helped boost my scientific research.

Finally, special thanks should go to my beloved family and in particular, my wife Xin Xu for their sustained support, encouragement, and loving considerations. I love you all!

Contents

Abbreviations

2D	Two-Dimensional
3D	Three-Dimensional
AFM	Atomic Force Microscope
ALD	Atomic Layer Deposition
ARPES	Angle-Resolved Photoemission Spectroscopy
CVD	Chemical Vapor Deposition
DFT	Density Function Theory
DOS	Density of States
EBH	Electron Beam Heating
ELF	Electron Localization Function
LEED	Low-Energy Electron Diffraction
MBE	Molecular-Beam Epitaxy
SEM	Scanning Electron Microscope
STEM	Scanning Transmission Electron Microscope
STM	Scanning Tunneling Microscope
TEM	Transmission Electron Microscope
TMDs	Transition Metal Dichalcogenides
UHV	Ultra-High Vacuum
XPS	X-ray Photoelectron Spectroscopy

Chapter 1
Introduction

Abstract This thesis focuses on the fabrication and physical properties study of novel two-dimensional (2D) atomic crystals beyond graphene. In the introduction, we first introduce readers the research background and then review four graphene-like 2D crystal materials, i.e., silicene, germanene, transition-metal dichalcogenides (TMDs), and hexagonal boron nitride (*h*-BN). Lastly, an introduction of research contents and instruments is presented.

Keywords Two-dimensional material · Atomic crystal · Graphene · Graphene-like · Van der Waals heterostructure

1.1 The Rise and Development of Graphene-Like 2D Crystal Materials

1.1.1 From Graphene to Graphene-Like 2D Crystal Materials

Graphene, a planar monolayer of carbon atoms with a two-dimensional (2D) honeycomb structure, is the first 2D atomic crystal. After its birth by isolation from graphite in 2004, it has become to be the hottest star in material science and attracted tremendous attraction in many different fields in the past ten years [1–3]. Its exotic properties, such as high surface area, high thermal conductivity, extremely high carrier mobility, and quantum Hall effect [4–9], enable potential graphene applications in electronics, photonics, energy generation and storage, and biosensors [10–14]. In order to achieve these promising applications, many preparation methods have been developed to produce few-to-monolayer graphene, such as mechanical and liquid-phase exfoliation, reduction of graphene oxides, chemical vapor deposition, and molecular beam deposition [15–23].

After an intensive study of graphene in the last ten years, "research on simple graphene has already passed its zenith" and "researchers have now started paying more attention to other two-dimensional atomic crystals such as isolated monolayers and few-layer crystals of hexagonal boron nitride (*h*-BN), molybdenum disulfide (MoS_2), other dichalcogenides and layered oxides", as claimed by Geim and Grigorieva [24]. Such materials present the entire range of electronic structures,

© Springer Nature Singapore Pte Ltd. 2020

L. Li, *Fabrication and Physical Properties of Novel Two-dimensional Crystal Materials Beyond Graphene: Germanene, Hafnene and PtSe*$_2$, Springer Theses,
https://doi.org/10.1007/978-981-15-1963-5_1

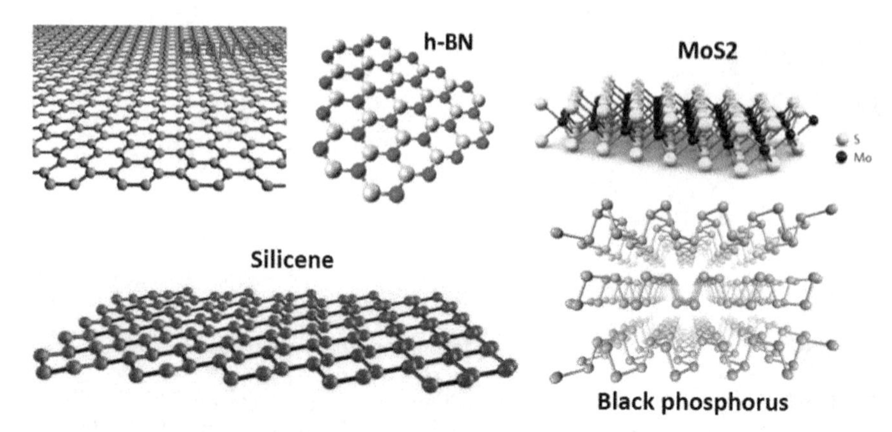

Fig. 1.1 Graphene and other two-dimensional materials. Reproduced with permission from Ref. [105], © 2014 Elsevier Ltd; [60], © 2015 Springer Nature; [59], © 2014, Springer Nature

from insulator to metal, from semiconductor to semi-metal and display many interesting properties, such as superconductivity, charge density wave, and topological insulator effect [25–46]. Moreover, graphene-like 2D buckled honeycomb structures have been attracting attention. These include silicene and germanene, the silicon- and germanium-based counterparts of graphene, which have been prepared experimentally after theoretical predictions and display atomic and electronic characteristics similar to graphene [47–54]. Based on their distinct properties, these 2D materials, as shown in Fig. 1.1, exhibit various application potentials on electronics, optoelectronics, catalysis, chemical sensors, and lithium-ion batteries [28–30, 39, 40, 42, 55–60].

1.1.2 Synthesis and Characterization

In general, all the methods employed for the synthesis and characterization of graphene could also be used effectively in the case of other 2D crystals. Thus, that is, single- and few-layer graphene-like 2D atomic crystals can be prepared by both physical and chemical methods. Mechanical exfoliation (scotch-tape technique), the first successful method for generating graphene, has already been used with other layered materials such as MoS_2 and $NbSe_2$ [61]. Liquid-phase cleavage assisted by ultrasonication has been employed to prepare many layered inorganic layered materials [62, 63]. Molecular beam epitaxial (MBE) has been successfully applied to the production of silicene and germanene [49, 53, 54]. Chemical methods include chemical synthesis, chemical vapor deposition (CVD), and atomic layer deposition (ALD). In particular, CVD is the most popular method for preparing large-scale 2D materials [35, 37, 64–68].

Single- and few-layer 2D materials are generally characterized by transmission electron microscope (TEM), scanning electron microscope (SEM), and scanning tunneling microscope (STM). Moreover, Atomic force microscope (AFM) has been demonstrated to be a powerful technique to determine layer thickness with a precision of 5%. Layer-dependent vibrational information can be obtained by Raman spectroscopy. X-ray diffraction can be employed to determine unit cell structure, the film thickness, and chemical constituents. Therefore, characterization of 2D materials is performed by a variety of microscopic and spectroscopic techniques.

1.2 Silicene and Germanene

1.2.1 Theoretical Investigations

The discovery and success of graphene have sparked new explorations on graphene-like films composed of other group-IV elements, such as so-called silicene and germanene, which are considered as Si- and Ge-based 2D counterparts of graphene. The planar honeycomb structure of graphene stems from the equivalently covalent bonds of carbon atoms formed by the fully sp^2-hybridized state. However, unlike graphene, in the case of Si and Ge, the sp^3-hybridized state is more stable than the planar sp^2-hybridized state. So that, silicene and germanene tend to form a mixed sp^2–sp^3 hybridized state with a buckled honeycomb structure rather than a planar one. That is also why there are no layered phases in bulk Si and Ge, in contrast to graphite (bulk C). Therefore, we cannot produce silicene or germanene sheets by micromechanical exfoliation of their bulk counterparts. However, we can still construct silicene and germanene structures by employing theoretical approaches. Actually, before exfoliation of graphene, first-principles calculations were performed to predict the most stable configurations for group-IV single sheets [69]. In contrast to the planar honeycomb lattice of graphene, Si and Ge monolayers can be stable with a nonplanar configuration, i.e., buckled honeycomb structure.

The first-principles calculations showed that the planar (PL) and high-buckled (HB) atomic structures of single-layer Si and Ge sheets are unstable, but the low-buckled (LB) honeycomb structures, as displayed in Fig. 1.2, can be stable, where the perpendicular distance between top and bottom Si (Ge) sublayers is $\Delta \approx 0.44$ Å (0.64 Å), forming the mixed sp^2–sp^3 hybridized orbitals [70]. In the LB honeycomb structure of silicene (germanene), the lattice constant and nearest-neighbor atomic distance are 3.86 Å (4.02 Å) and 2.28 Å (2.42 Å), respectively. In contrast to the C–C distance (1.42 Å) in graphene, much larger Si–Si and Ge–Ge atomic distances severely decrease the electronic overlaps of π–π bonds.

S. Cahangirov et al. calculated the electronic energy band spectra of Si and Ge for HB, PL and LB configurations and corresponding density of states (DOS) for LB structures [70]. As shown in Fig. 1.3, both HB Si and Ge are metallic. Similar to graphene, π and π^* bands of PL and LB silicene cross at the K point in the Fermi

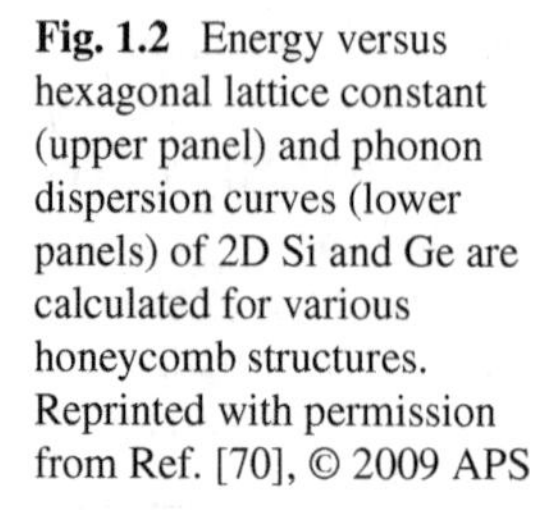

Fig. 1.2 Energy versus hexagonal lattice constant (upper panel) and phonon dispersion curves (lower panels) of 2D Si and Ge are calculated for various honeycomb structures. Reprinted with permission from Ref. [70], © 2009 APS

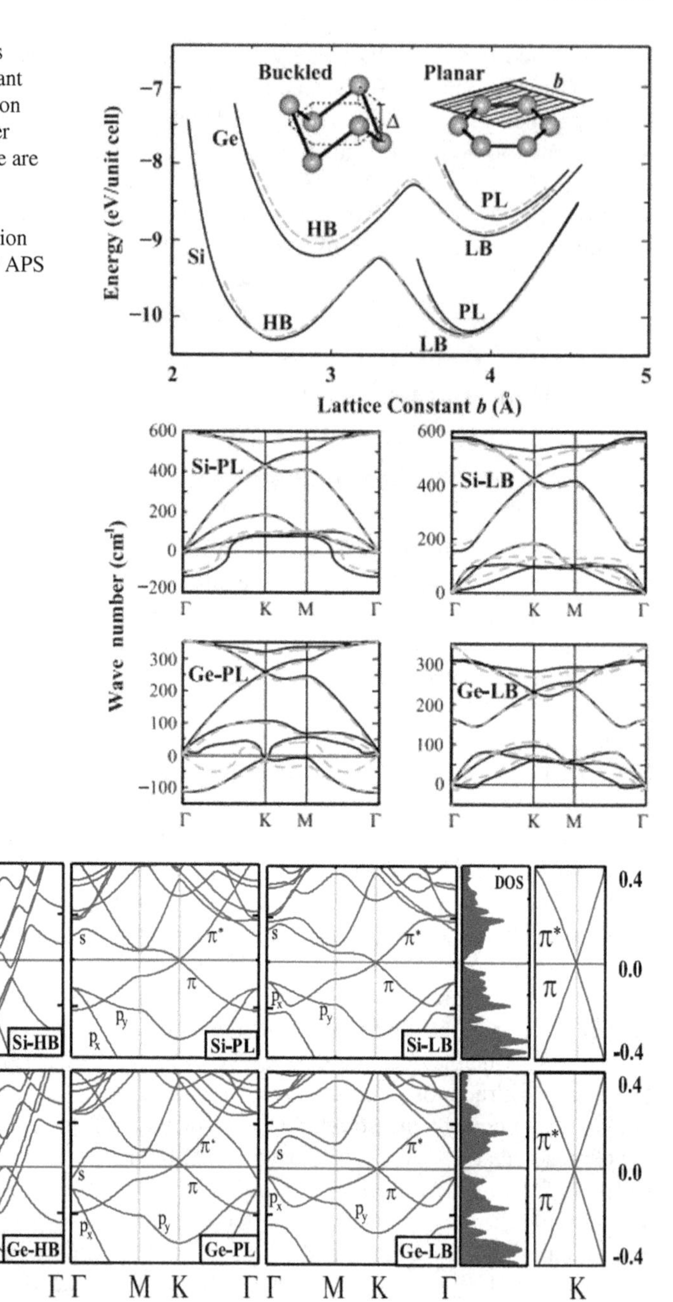

Fig. 1.3 Band structures of Si and Ge are calculated for high-buckled (HB), planar (PL), and low-buckled (LB) structures. For LB structure the density of states (DOS) is also presented. Reproduced with permission from Ref. [70], © 2009 APS

level (EF) and thus they are semimetallic. Planar Ge structure is metallic due to a low density of s-like states crossing the EF. In contrast, LB structure of Ge shows semimetallic behavior with linear π and π^* bands crossing at K point in the EF. This behavior of bands is attributed to a mass-less Dirac fermion character of the charge carriers.

1.2.2 Experimental Explorations

Theoretical investigations demonstrated that silicene (germanene) is a metastable structure, with 0.79 eV/atom (0.67 eV/atom) higher energy compared to bulk Si (Ge) structure. It suggests thickness-controllable deposition technique is preferred for silicene and germanene growth, such as molecular beam epitaxy and low-temperature atomic layer deposition, in order to avoid the formation of three-dimensional structures. It should be mentioned that there is not yet any report on the successful preparation of germanene prior to our research (the early explorations of germanene growth are exhibited in Sect. 2.1.2). Therefore, we will only review the experimental investigations on the growth of silicene as follows.

Experimental explorations of silicene began with silver (Ag) surfaces, as Si/Ag system tends to phase separation due to the very low solubility of Si in bulk Ag.

First, silicon nanoribbons were obtained by Si deposition onto Ag(110) surface under ultrahigh vacuum [71–73]. STM measurements revealed Si honeycomb structures inside the nanoribbon, as shown in Fig. 1.4. Correspondingly, DFT calculations showed that Si atoms have a tendency toward honeycomb arrangement on the Ag surfaces. Moreover, the simulated STM images displayed a buckled honeycomb structure, in agreement with the experimental observations. This is the first evidence of the existence of silicene.

In 2010, Lalmi, B. et al. claimed that they succeeded in preparing a continuous, two-dimensional sheet of silicene on Ag(111) surface [74]. However, the Si–Si atomic distance in the honeycomb structure is only 0.19 nm directly measured

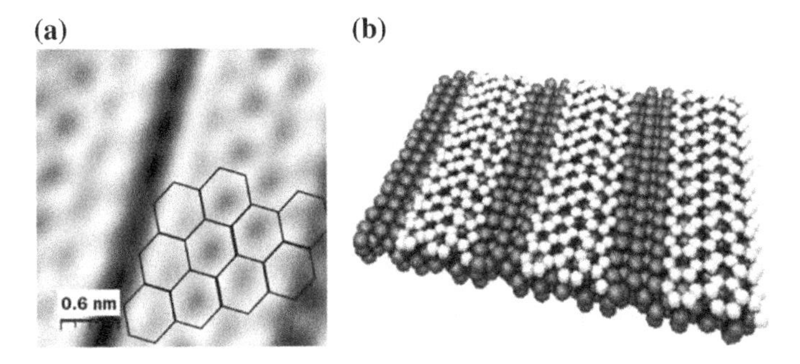

Fig. 1.4 **a** High-resolution STM image and **b** calculated atomic structure of silicon nanoribbons. Reprinted from Ref. [71], © 2010 AIP Publishing; [72], © 2010 AIP Publishing

from their STM images, which is too much smaller than the calculated value (0.22–0.24 nm). Le Lay et al. argued that the observed honeycomb lattice did not correspond to a silicene layer but instead to the bare Ag(111) surface with a contrast inversion [75]. In subsequent studies of the growth of silicene on Ag(111) surface, the 4×4 and $(\sqrt{13} \times \sqrt{13})R13.9°$ superstructures have been found [76] and a $(\sqrt{7} \times \sqrt{7})R19.1°$ structure has been predicted [77], as shown in Fig. 1.5. Feng et al. also found the 4×4 and $(2\sqrt{3} \times 2\sqrt{3})R30°$ superstructures, although they proposed different theoretical structural models [49]. More importantly, they revealed a new superlattice $(\sqrt{3} \times \sqrt{3})R30°$, as illustrated in Fig. 1.6. They constructed the following atomic structure: two Si atoms in a honeycomb ring are buckled upward and one atom is buckled downward, forming a tri-sublayer structure. They also studied the dI/dV spectra of silicene and found the Dirac point of silicene at around 0.52 ± 0.02 eV.

Our group has been involved in the research of graphene and graphene-like 2D materials for years. Recently we successfully prepared silicene sheets on an Ir(111) surface by MBE [53]. STM and LEED observations, as presented in Fig. 1.7, revealed that the silicene layer possesses a $(\sqrt{7} \times \sqrt{7})$ superstructure with respect to the underlying Ir(111) surface, which agrees well with a low-buckled atomistic model of silicene generated by DFT calculations. Importantly, the calculated electron localization function revealed 2D continuity of silicene layer on the Ir(111) substrate. This work provides a method to fabricate high-quality silicene and an explanation for the formation of the buckled silicene sheet.

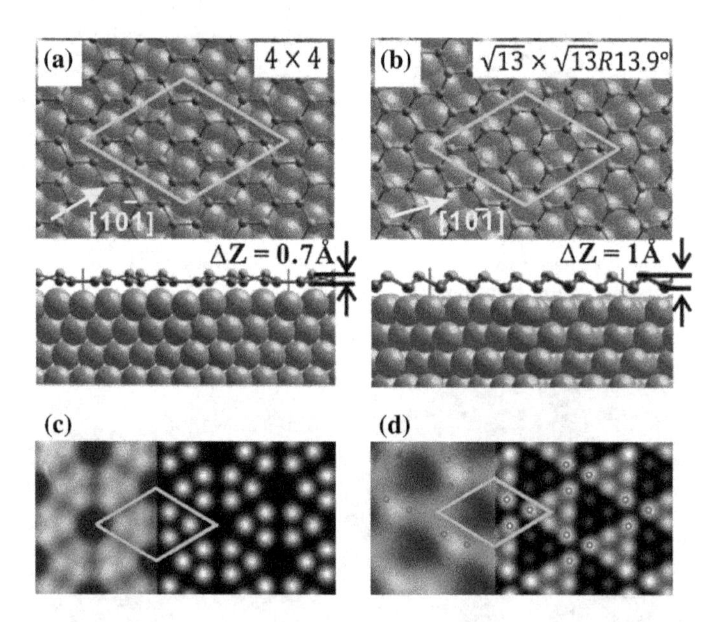

Fig. 1.5 Optimized structure models and simulated STM images of **a**, **c** 4×4 and **b**, **d** $(\sqrt{13} \times \sqrt{13})R13.9°$ silicene on Ag(111). Reproduced with permission from Ref. [76], © 2012 The Japan Society of Applied Physics

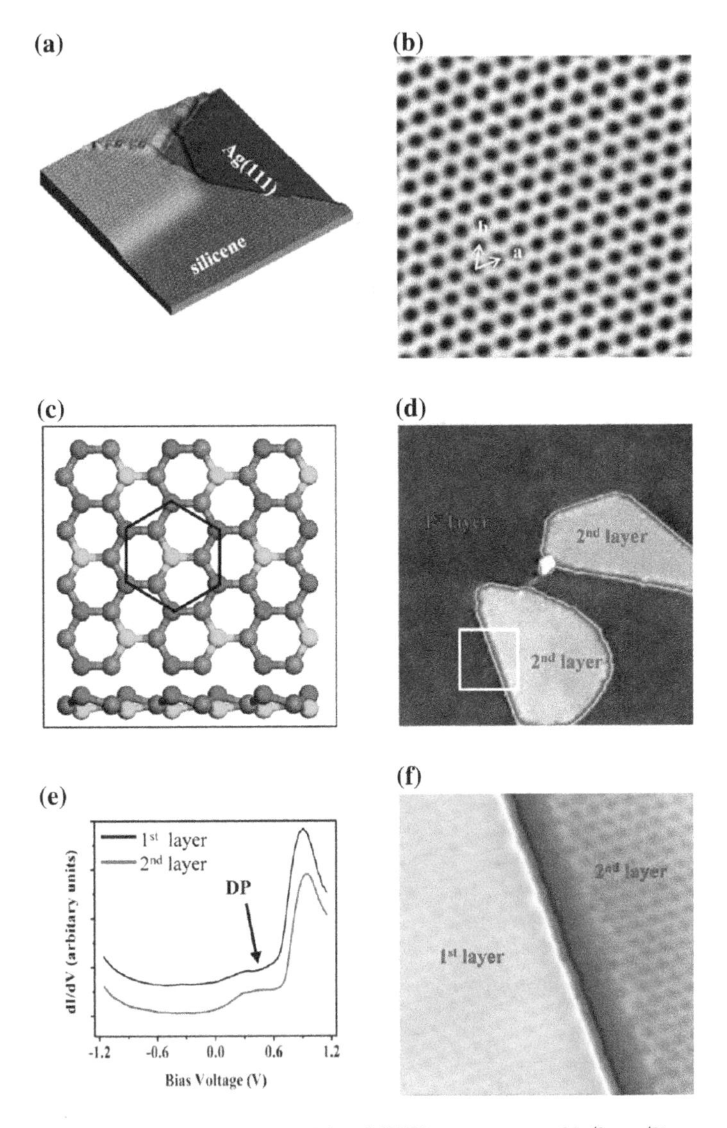

Fig. 1.6 STM observations, schematic model, and dI/dV spectroscopy of ($\sqrt{3} \times \sqrt{3}$) superstructure of silicene on Ag(111). Reprinted with permission from Ref. [49], © 2012 ACS

Besides the metal substrates mentioned above, there is a report on the growth of silicene on ZrB_2 surface supported on a Si(111) substrate [50], as shown in Fig. 1.8. Instead of directly depositing silicene, the researchers found a (2×2) reconstructed structure on the ZrB_2(0001) surface, which was considered to be separated from the underlying Si(111) substrate and assigned to the ($\sqrt{3} \times \sqrt{3}$) superstructure of silicene layer. They characterized and studied its atomic structures and electronic properties by performing STM, XPS, ARPES, and ab initio calculations.

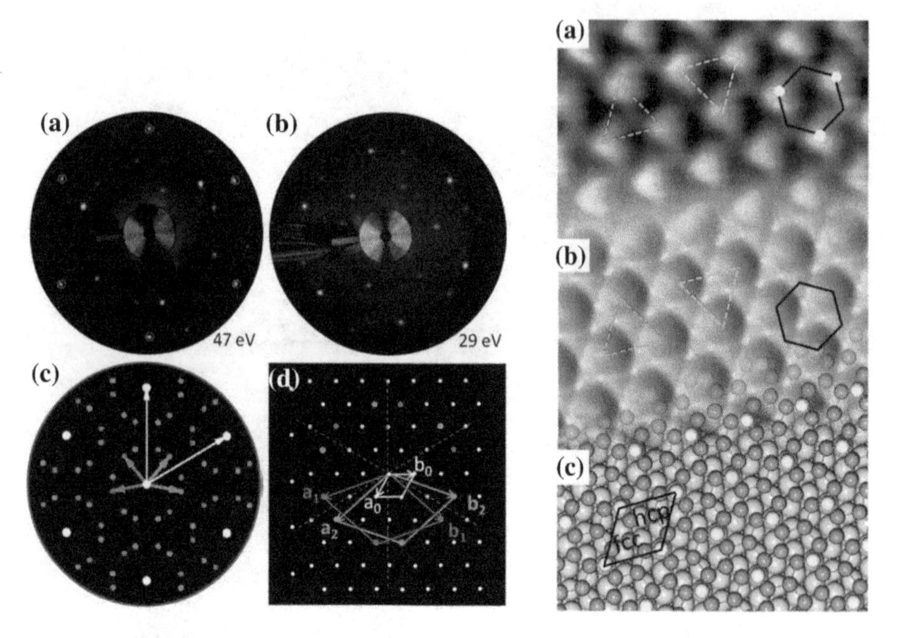

Fig. 1.7 Left: **a**, **b** LEED patterns and **c**, **d** theoretical LEED patterns of ($\sqrt{7} \times \sqrt{7}$) superstructure of silicene. Right: **a** STM image, **b** simulated STM image, and **c** calculated atomic structure of ($\sqrt{7} \times \sqrt{7}$) superstructure of silicene on Ir(111). Reproduced with permission from Ref. [53], © 2013 ACS

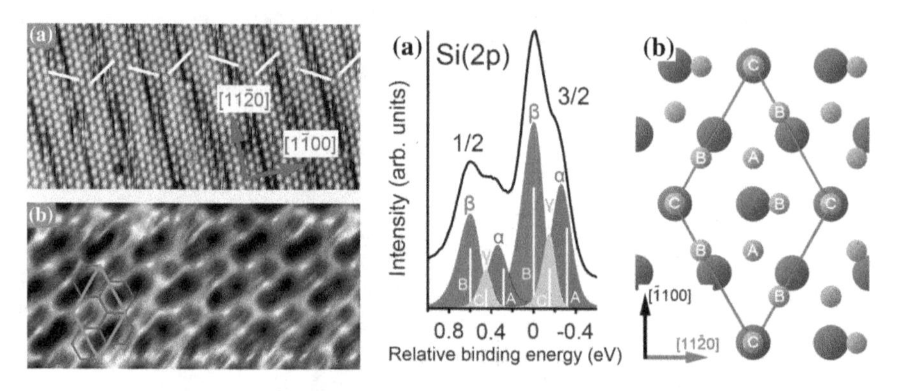

Fig. 1.8 High-resolution STM images, XPS measurement, and atomic model of Si (2 × 2) reconstructed structure on $ZrB_2(0001)$ surface. Reproduced from Ref. [50]

1.3 Transition Metal Dichalcogenides

1.3.1 *The Rise of TMDs*

In the past ten years, the intensive studies of graphene and the rapid progress in the methodology developed in preparing ultrathin films have sparked new explorations and discoveries of novel 2D crystal materials beyond graphene. In particular, layered transition metal dichalcogenides (TMDs) have attracted considerable attention because of their diverse physical properties and a variety of potential applications [30, 58]. It is well known that graphene, on the one hand, is chemically inert and can only be functionalized by introducing desired molecules, which in turn will influence its novel properties. On the other hand, graphene has no bandgap, which considerably weakens and hinders its technological applications in the semiconductor industry. Band gaps can be opened with experimental methods, such as chemical functionalization, fabrication of graphene nanostructures, but these methods increase the process difficulty and result in the loss of electron mobility.

In contrast, TMDs show a wide range of physical and chemical properties, with their electronic structures ranging from insulators, semiconductors, to semimetals, and pure metals. Some bulk TMDs materials show low-temperature phenomena, such as charge density wave (CDW) and superconductivity [25, 31]. The exfoliated few- or single-layer TMDs nanosheets exhibit properties distinct from that in graphene [28, 29, 32], which could complement and enhance the application potentials of graphene. For example, several monolayer TMDs possess sizable bandgaps around 1–2 eV, which enable them to be promising optoelectronic devices [26, 30, 40, 55, 56]. The versatile properties of TMDs offer a platform for both fundamental research and technological applications.

1.3.2 *Structural Properties of TMDs*

The formula of TMDs materials is MX_2, where M is one of transition metal elements of groups 4–10 (such as Hf, Ta, Mo, W, and Pt), and X is a chalcogen (S, Se or Te). These layered crystals adopt sandwich-like structures with the form X-M-X, where a sublayer of metal atoms is sandwiched by two hexagonally packed planes of chalcogen atoms. An individual MX_2 monolayer is defined by three sublayers of atoms (X-M-X) in which the M and X atoms are bonded covalently. Adjacent monolayers are weakly held by Van der Waals forces to form 3D stacked crystal, as depicted in Fig. 1.9. Bulk TMD crystals possess three polymorphs depending on stacking orders and the coordination of metal atoms by the chalcogens. In both 2H and 3R phases, the metal coordination is trigonal prismatic (honeycombs), while the coordination in 1T-MX_2 is octahedral (centered honeycombs). Here, the digit indicates the number of X-M-X layers in each stacking unit cell, and the letters represent the types of symmetry: hexagonal (H), rhombohedral (R), and trigonal (T). For instance, bulk MoS_2

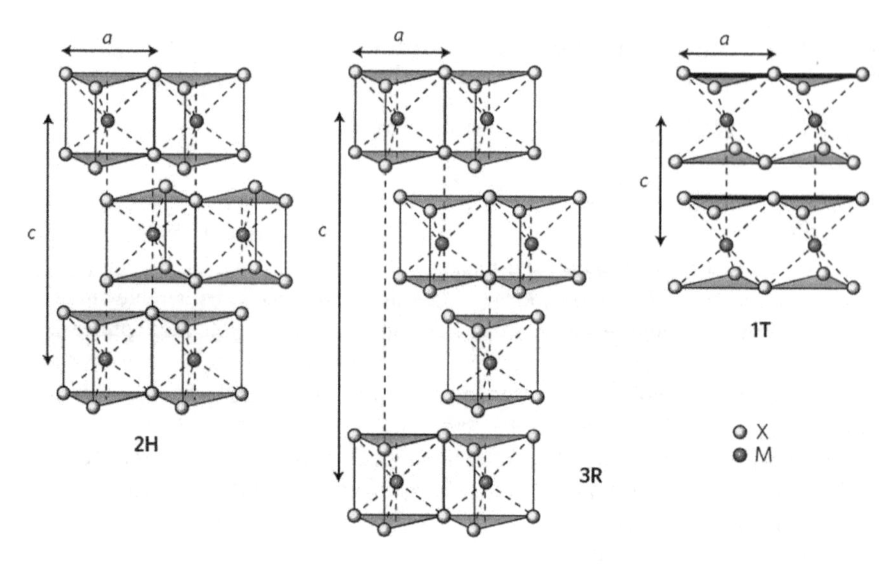

Fig. 1.9 Schematics of the structural polytypes of layered TMDs materials. Reproduced with permission from Ref. [30], © 2012 Springer Nature

is a prototypical TMD commonly found in the "2H phase", in which each MoS_2 unit cell contains two layers of S-Mo-S sandwiches with a hexagonal symmetry in the vertical projection.

Note that monolayer TMDs show only two polytypes: trigonal prismatic and octahedral phases, generally referred to as 1H and 1T MX_2, respectively, as depicted in Fig. 1.10.

1.3.3 Electronic Properties of TMDs

1.3.3.1 Electronic Properties of Bulk TMDs

Depending on the coordination and oxidation state of the metal atoms, layered TMDs can be semiconducting (e.g. M = Mo, W) or metallic (e.g. M = Ta, Nb). As shown in Fig. 1.11, a simple model for ideal coordination is exhibited to explain the diverse electronic properties of TMDs crystals. In both 1H and 1T phases, the non-bonding d orbitals of the TMDs are located within the bandgap of bonding (σ) and anti-bonding states (σ*) in group 4, 5, 6, 7 and 10 TMDs. According to ligand field theory, Octahedrally coordinated transition metal centers of TMDs form two non-bonding d orbitals, $d_{yz,xz,xy}$ (bottom) and $d_{z2,\ x2-y2}$ (top), while transition metals with trigonal prismatic coordination exhibit three d orbitals, d_{z2}, $d_{x2-y2,xy,}$ and $d_{xz,yz}$ (from bottom to top). The diversity of electronic properties of TMDs arises from the progressive filling of the non-bonding d orbitals from group 4 to group 10 species. When an

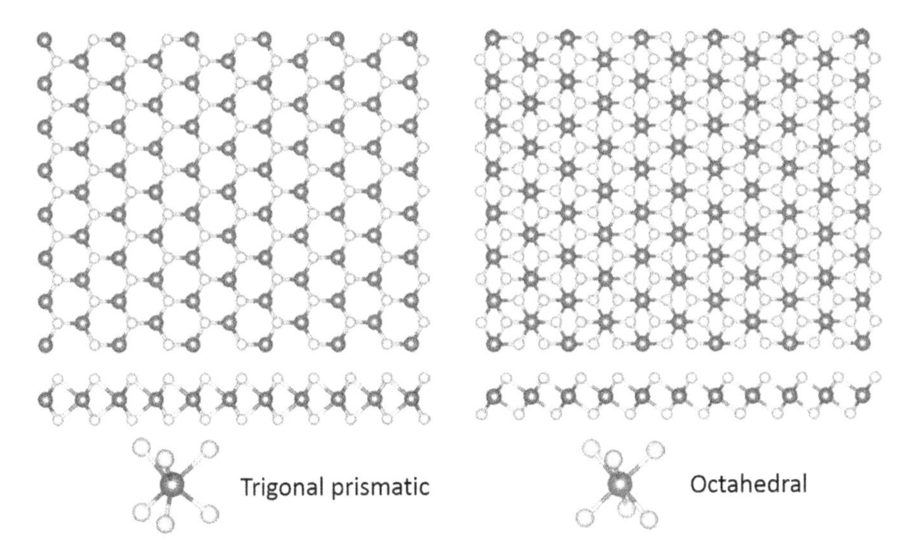

Fig. 1.10 Structural models of single-layer TMDs with trigonal prismatic and octahedral coordination. Reprinted with permission from Ref. [58], © 2013 Springer Nature

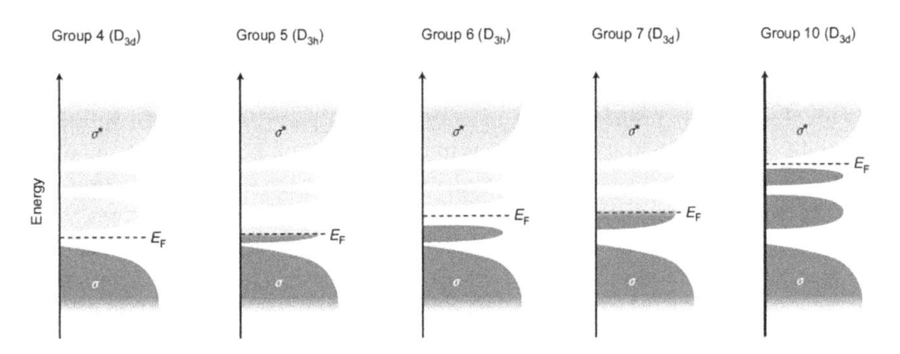

Fig. 1.11 Schematic illustration showing progressive filling of d orbitals of TMDs. Reproduced with permission from Ref. [58], © 2013 Springer Nature

orbital is partially filled, as in the case of 2H-$NbSe_2$ and 1T-ReS_2, the Fermi level is within the band, and the compounds exhibit a metallic character. When an orbital is fully filled, such as in 1T-HfS_2, 2H-MoS_2, and 1T-PtS_2, the Fermi level is located in the energy gap, and a semiconducting character is observed.

1.3.3.2 Band Structures of Single-Layer TMDs

Due to the effect of quantum confinement and changes in interlayer coupling, single-layer TMDs prepared by deposition or isolation go through considerable changes in

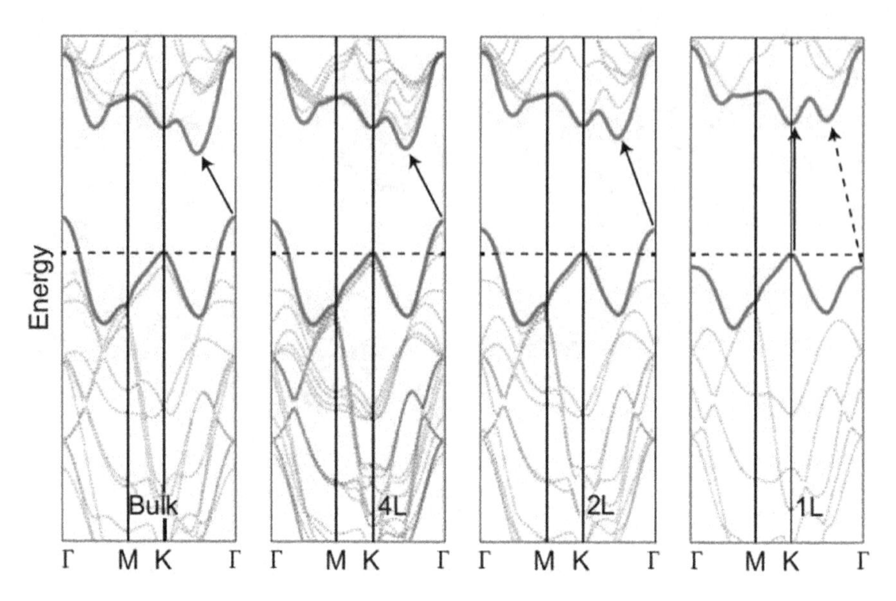

Fig. 1.12 Band structures of bulk and few- to mono-layer MoS_2. Reprinted with permission from Ref. [55], © 2010 ACS

band structures compared to their bulk counterparts. For example, the band structures of bulk, few- and mono-layer MoS_2 calculated from first principles are exhibited in Fig. 1.12. At the Γ point, the bandgap is indirect for the bulk material but gradually shifts to be direct for the monolayer. This indirect-to-direct bandgap transition with layer number is due to quantum confinement and the resulting change in hybridization between p_z orbitals on S atoms and d orbitals on Mo atoms. Several studies have confirmed a similar band transition for $MoSe_2$, WS_2, and WSe_2, which accounts for the enhanced photoluminescence in monolayers of MoS_2, $MoSe_2$, WS_2, and WSe_2, in contrast to only weak emission observed in multilayered forms [26, 55, 56, 78, 79]. The transition to a direct bandgap in the monolayer form enhances the absorption and emission efficiency of photons and thus has important applications for the photonics, optoelectronics, and sensing.

1.3.4 Synthesis of TMDs

Preparation of materials is the prerequisite for fundamental research, and the essential step toward translating their unique properties into applications. Until now, many methods have been employed to prepare ultrathin TMDs materials. The top-down methods, which rely on the exfoliation of layered bulk crystals, include the mechanical cleavage method, liquid-phase exfoliation by direct sonication, chemical Li-intercalation and exfoliation, and laser thinning technique. Examples of bottom-up approaches are CVD growth and wet chemical synthesis.

1.3.4.1 Top-Down Preparation

Mechanical exfoliation, as shown in Fig. 1.13, is the original technique developed for graphene [80]. It remains the best method for preparing high-purity atomically thin flakes of TMDs peeled from their layered TMD bulks with a view to investigating their inherent physical properties and fabrication of individual devices [61]. However, this method is not scalable and cannot allow precise control of film thickness and size. Therefore, this method cannot be used for mass production in industrial applications.

Liquid-phase exfoliation has been developed to obtain large quantities of exfoliated ultrathin flakes [62, 63, 81–84], which involves dissolution and exfoliation of TMDs powders by direct sonication in commonly used solvents such as dimethylformamide and N-methyl-2-pyrrolidone. One of the disadvantages of liquid exfoliation is the difficulty in preparing monolayer TMD sheets and maintaining the lateral size of exfoliated nanosheets. Ultrasonic-promoted hydration of lithium-intercalated compounds, as illustrated in Fig. 1.14, is another effective method for mass production of fully exfoliated TMDs layers. This approach utilizes a solution of a lithium-containing compound such as n-butyllithium to achieve intercalation of lithium ions between the layers and thus rapidly separates them into single layers. Although the yield of the lithium intercalation method for obtaining single-layer TMDs is nearly 100%, some challenges remain. The lithium intercalation must be carefully controlled to obtain complete exfoliation instead of the formation of metallic compounds, such as Li_2S. Otherwise, the resulting exfoliated material differs structurally and electronically from the bulk material due to the charge transfer between TMDs and alkali ions.

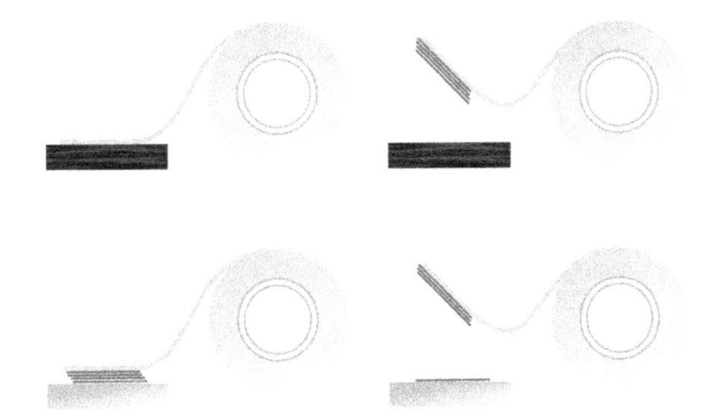

Fig. 1.13 Illustrations of the "Scotch-tape" method of producing graphene. Reproduced with permission from Ref. [80], © 2011 APS

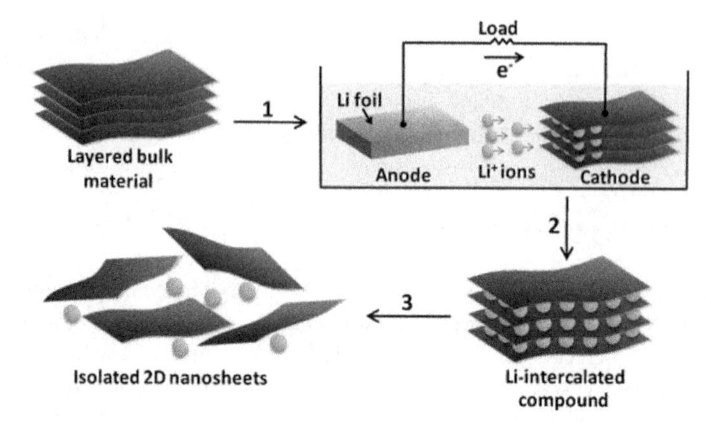

Fig. 1.14 Electrochemical lithiation process for the fabrication of 2D nanosheets from the layered bulk material. Reproduced with permission from Ref. [82], © 2011 Wiley

1.3.4.2 Bottom-up Synthesis

Reliably synthesizing high-quality and large-area ultrathin films is an essential step for applications in electronics and optoelectronics. For example, chemical vapor deposition of graphene on copper foils, as a significant breakthrough in the preparation of large-area graphene, has enabled large-scale device fabrication. Various CVD and CVD-related methods have been developed for growing atomically thin TMDs films [65, 66, 85, 86], as shown in Fig. 1.15, which include three strategies summarized as follows. (I) vaporization of metal and chalcogen precursors and their decomposition, followed by deposition of the resulting TMDs on a substrate, (II)

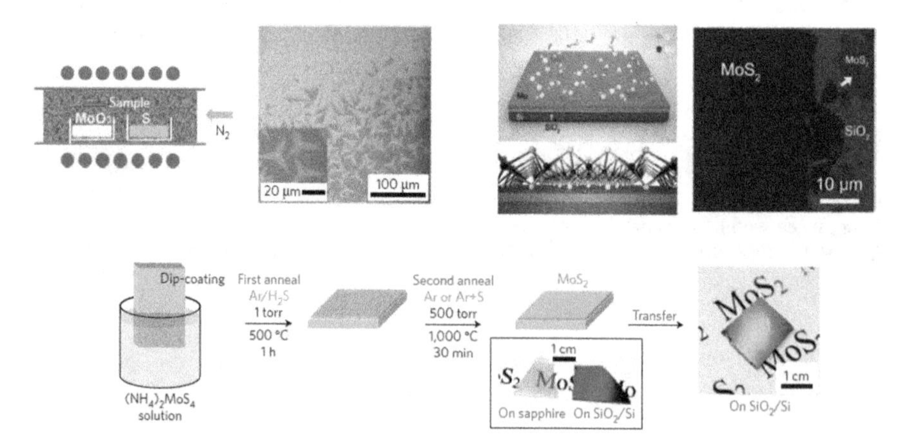

Fig. 1.15 Schematics of CVD methods for synthesizing MoS_2. Reprinted with permission from Ref. [85], © 2012 Wiley; [66], © 2012 Wiley; [65], © 2012 ACS

direct sulfurization of the pre-deposited metal film, and (III) conversion of transition metal oxide (such as MO_3) to TMDs (MX_2) by sulfurization or selenization.

Besides CVD methods, chemical syntheses of WS_2, MoS_2, WSe_2, and $MoSe_2$ have been demonstrated using hydrothermal synthesis [87–90].

1.3.5 Applications of TMDs

The unique electronic properties of TMDs, especially semiconducting TMDs enable them to have many significant applications for electronics, optoelectronics, energy storage, photovoltaic cells, and photoelectric detection, as will be discussed in more detail below.

1.3.5.1 Electronics

One of the most extensive and important applications of semiconductors is the fabrication of digital circuit transistors. Nowadays, the size of silicon-based field-effect transistors (FETs) has approached its physical limit, motivating people to explore new alternative materials. Digital logic transistors require the properties of high charge carrier mobility, a large on/off current ratio, high conductivity, and low off-state conductivity. Graphene has aroused intense interest due to its two-dimensional characteristics and ultra-high electron mobility [13, 91–93]. However, the lack of energy gap means that the transistor made of graphene cannot achieve low off-state current and thus brings about energy consumption, nor can it effectively control switching with high on/off ratios. Now more and more attention has been paid to ultra-thin semiconducting TMDs, as displayed in Fig. 1.16. As promising field-effect transistor channel materials, TMDs show stable structures free of surface dangling bonds and comparable mobility to those of Si. Their atomic-scale thickness combined with the bandgaps in the range of 1–2 eV enables large on/off switching ratios, a high degree of immunity to short channel effects and thus a considerable reduction in power dissipation [57, 94, 95].

1.3.5.2 Optoelectronics

The electronic band structures of semiconductors play an essential role in optical absorption and emission. For indirect-bandgap semiconductors, they need an additional process of phonon absorption or emission to satisfy the momentum conservation, thus reducing the efficiency of photon absorption and emission. In contrast, many TMDs monolayers, such as MX_2 ($M = Mo, W, X = Se, Te$), possess direct semiconducting bandgaps, in which photons with energy larger than the bandgap width can be readily absorbed and emitted. In addition, because of their sub-nanoscale thinness and processability, they have sparked considerable interest in a wide range

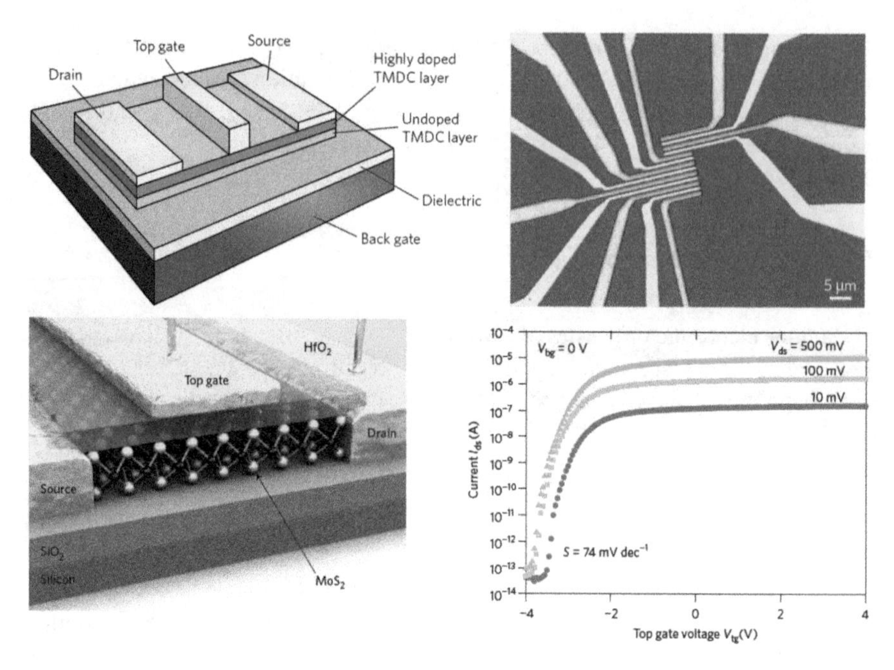

Fig. 1.16 Schematic illustrations of TMDs-based transistors. Reprinted with permission from Ref. [30], © 2012 Springer Nature; [57], © 2011 Springer Nature

of potential applications for flexible and transparent optoelectronic devices [27, 96], as shown in Fig. 1.17.

1.3.5.3 Light Emission

Light emission can be classified into photoluminescence (PL) and electroluminescence (EL). PL is light emission from a material after the absorption of photons (electromagnetic radiation). It is initiated by photoexcitation (i.e., photons that excite electrons to a higher energy level in an atom). Following excitation, various relaxation processes typically occur in which other photons are re-radiated. PL process is observed in monolayer MoS_2, which has a direct bandgap, and the photoluminescent efficiency is demonstrated to be much higher than those of bilayer and bulk MoS_2, both of which are indirect-bandgap semiconductors [55, 56]. EL is an optical phenomenon and electrical phenomenon in which a material emits light in response to the passage of an electric current or to a strong electric field. It is the result of radiative recombination of electrons and holes in a material, usually a semiconductor. The excited electrons release their energy as photons-light [97, 98]. In semiconductors with direct band gaps, the combination of electrons and holes and photonic radiation occur more efficiently than in indirect bandgap semiconductors. So the

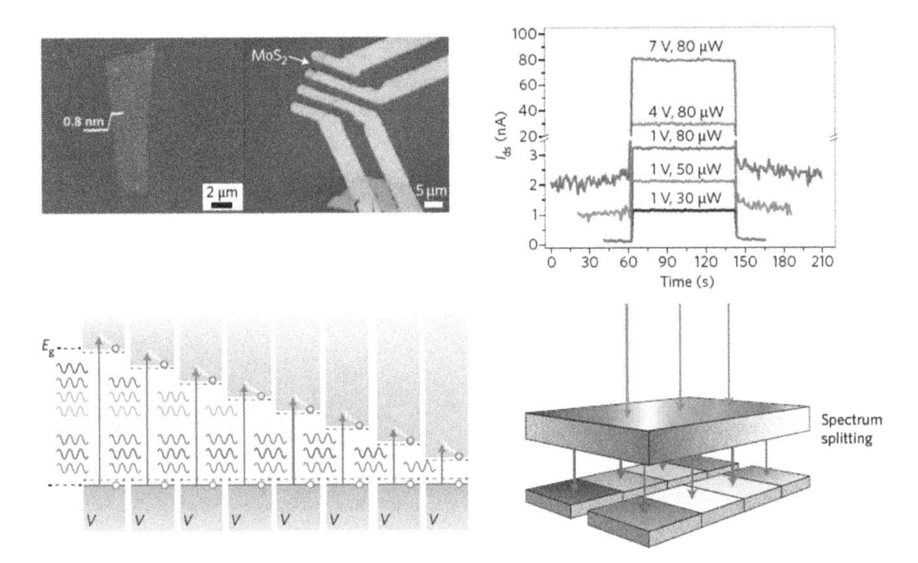

Fig. 1.17 Current and proposed TMDs optoelectronic devices. Reproduced with permission from Ref. [27], © 2011 ACS; [96], © 2012 Springer Nature

direct bandgaps of monolayer semiconducting TMDs make them the promising light emission materials in flexible optoelectronic devices.

1.4 *h*-BN and Other Graphene-Like 2D Materials

Bulk *h*-BN, called "white graphite", has a similar layered structure as graphite with the exception that the basal planes in *h*-BN are vertically aligned to each other, with the electron-deficient B atoms in one layer lying over and above the electron-rich N atoms in adjacent layers, as illustrated in Fig. 1.18. For graphite, however, the adjacent layers are stacked offset, and alternating C atoms lie above and beneath the center of the honeycomb structures. Both bulk *h*-BN and graphite exhibit very similar lattice constants and interlayer distances. Due to their close structural similarity, single-layer *h*-BN can thus be regarded as a structural and isoelectronic analog of graphene, composed of alternating boron and nitrogen atoms in a honeycomb arrangement. It can be obtained by replacing C–C bond with B–N bond in graphene [99, 100].

Although *h*-BN and graphene have similar structures, their electronic properties are strikingly distinct. The pristine *h*-BN sheets are intrinsically insulators or wide band-gap semiconductors (approximately 5.9 eV) [101], in contrast to semi-metallic graphene. Because of its excellent electrical insulation property, *h*-BN has been applied as a charge leakage barrier layer for use in electronic equipments. On the other hand, 2D *h*-BN has many other excellent properties. For example, *h*-BN has

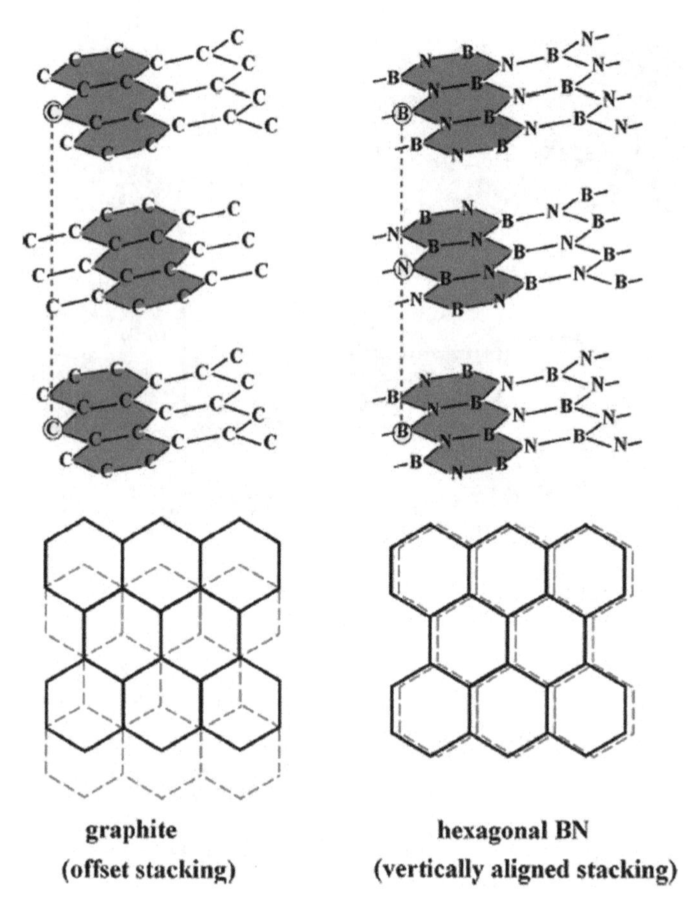

Fig. 1.18 Structures of graphite and *h*-BN. Reproduced with permission from Ref. [100], © 2013 Elsevier Ltd

excellent thermal conductivity and mechanical strength, good optical properties than graphene, and better chemical and thermal stability.

With a similar lattice constant to graphene and the same hexagonal structure, *h*-BN layers, as ideal planar insulating substrates, offer one of most advanced platforms for graphene-based electronics by enhancing the stability, quality and carrier mobility of graphene [102]. For example, *h*-BN sheets can be used as good back-gate dielectric layer materials for graphene-based field-effect transistors. Because of their commensurate structural parameters and distinct electronic properties, layered heterostructures consisting of graphene and *h*-BN layers have recently attracted intense interests. Stacking these layers in a precisely controlled sequence can give rise to new phenomena and create tailored properties [103, 104].

Aside from *h*-BN, many other 2D materials have also earned tremendous attention, such as graphene derivate (e.g. hydrogenated graphene, fluorinated graphene), layered group-IV and group-III metal chalcogenides (e.g. SnX (X = S, Se, or Te)), layered oxides and hydroxides (e.g. V_2O_5, ZnO). The investigations into these novel materials will generate fantastic properties and amazing applications.

1.5 Van der Waals Heterostructures

From the above sections, we can see that since the discovery of the exotic properties of graphene, graphene-like 2D materials have gained widespread attention due to their unique properties that could have many fundamental applications. The ideas and methodology developed in preparing graphene, such as mechanical peeling, molecular beam epitaxial and chemical vapor deposition, have been extended to the exploration of other 2D materials, including some artificially synthesized atomic crystals that do not exist in nature. Moreover, we can tune the structures and electronic properties of these two-dimensional materials by various methods, such as chemical or molecular doping, atomic or molecular intercalation, and exertion of tension or pressure. It is not difficult to imagine that we can pile up these 2D layers in a controlled manner to form new 3D stacked materials, as shown in Fig. 1.19. Given that each individual component exhibits distinct properties, we possess a library of 2D atomic crystals, which allows us to create 3D layered structures with novel and tailored properties, with an ultimate desire for a vast library containing various functional materials. This is the basic principle of van der Waals heterostructures techniques, a growing research field in the past years [24, 105]. This approach involves three

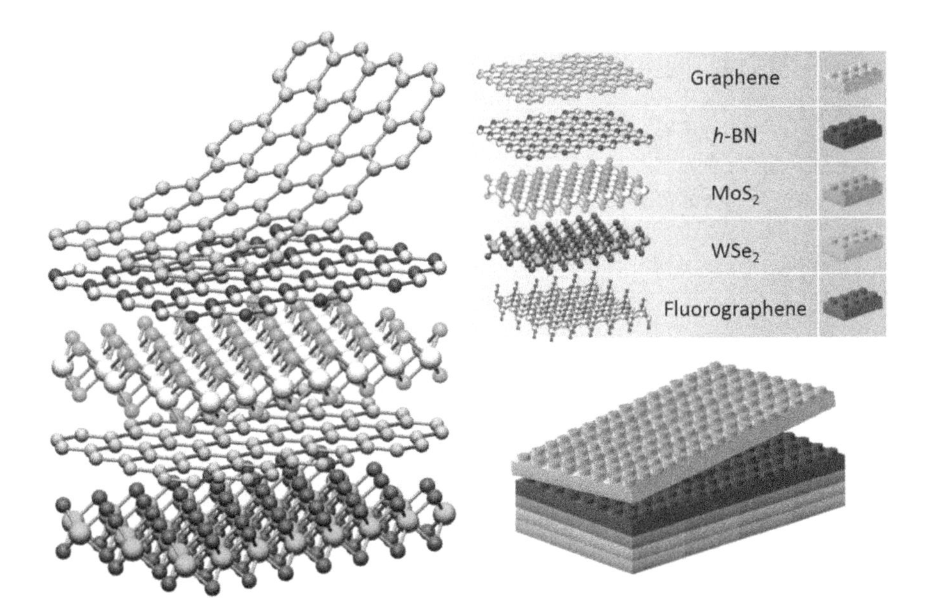

Fig. 1.19 Van der Waals heterostructures. Reprinted from Ref. [24], © 2013 Springer Nature

processes: (a) preparing various 2D materials by mechanical exfoliation and/or synthetic growth; (b) tuning electronic, optical and chemical properties of these 2D materials by chemical modification or applying stress or external field; (c) stacking these 2D crystals on top of each other in a controlled fashion to generate new hybrid structures. The development of these novel heterostructures paves a way towards a variety of practical applications since it could create artificial materials that exhibit unique properties and multiple functions.

1.6 Research Contents and Instruments

1.6.1 Research Contents

The successful exfoliation of graphene out of graphite resulted in the first real two-dimensional material in nature. Its unique electronic structures and exotic properties were subsequently revealed, followed by fruitful results of graphene-related fundamental researches and practical applications. Graphene is undoubtedly the most dazzling research focus in the field of material science in the past decade. More importantly, graphene opens up, or at least reactivates the research field of 2D materials, and arouses researchers' enthusiasm for the exploration of 2D materials. A focused effort on growth and isolation of high-quality single-layer nanosheets was re-initiated after the discovery of graphene, including some two-dimensional systems that have been studied in the past and re-examined now from a new perspective. In this context, the research in this thesis was performed on graphene-like 2D crystal materials, focusing mainly on germanene, hafnium honeycomb crystal, and platinum diselenide. The preparation methods, atomic structures, electronic properties as well as practical applications of the above materials were studied systematically.

The thesis contains five parts and is ordered as follows. Following this introduction part, Chap. 2 presents the growth method of single-layer germanene and reveals its superstructures on the metal substrate by means of experimental characterizations and theoretical calculations. In Chap. 3, the fabrication of honeycomb-like structures based on the transition metal element is reported for the first time. Atomically-resolved STM images show that hafnium atoms form a long-range ordered honeycomb structure on an Ir (111) substrate. The calculated density of states and STM simulation confirm that Hf atoms are assembled into a continuous 2D honeycomb lattice through covalent bonding. Chap. 4 involves another type of 2D layered materials beyond graphene-transition metal dichalcogenides (TMDs). We report on the preparation of a new TMD, monolayer platinum diselenide ($PtSe_2$) on a Pt(111) surface, by a brand new synthesis strategy. A variety of surface characterization methods have revealed the atomic structure and interfacial characteristics of single-layer $PtSe_2$. Angle-resolved photoemission spectroscopy (ARPES) demonstrates its semiconducting features. Besides, we have carried out photocatalytic investigations and predicted the application in valley electronics theoretically. Lastly, the thesis ends in the 5th chapter with important conclusions and prospect of the work.

1.6.2 *Introduction of Instruments*

Most of the experiments involved in this thesis were performed on the ultra-high vacuum (UHV) chambers equipped with a commercial variable-temperature scanning tunneling microscope (Omicron VT-STM), as shown in Fig. 1.20. The system contains a preparation chamber, an analysis chamber, a load-lock chamber for rapid sample transmission and other ancillary facilities. The preparation chamber is the largest chamber, which includes many instruments for substrate cleaning, sample preparation, and experimental characterization, such as Ar ion sputtering gun, low-energy electron diffraction (LEED), electron-beam heating stage (EBH), and molecular beam epitaxy (MBE) evaporators. All the single crystals are cleaned by several cycles of sputtering with the Ar ion sputtering gun and annealing on the EBH. The clean and flat metal surfaces serve as the epitaxial substrates for growth of 2D materials. The growth process and the structural properties are controlled using the EBH by accurate control over the growth temperature. LEED is used for the determination of the surface structure and symmetry of 2D atomic crystals. The leakage valves connected to various external gases provide a precisely controllable gas pressure for the synthesis of 2D films. These instruments and facilities on the preparation chamber ensure the controllable fabrication of 2D crystalline materials. In addition, the preparation chamber is connected to a small load-lock chamber, which allows us to load and transfer samples from the atmosphere in a fast manner without breaking the vacuum of the preparation chamber. The core part of the whole system is the VT-STM in the analysis chamber. It can obtain atomic-resolution topographic and

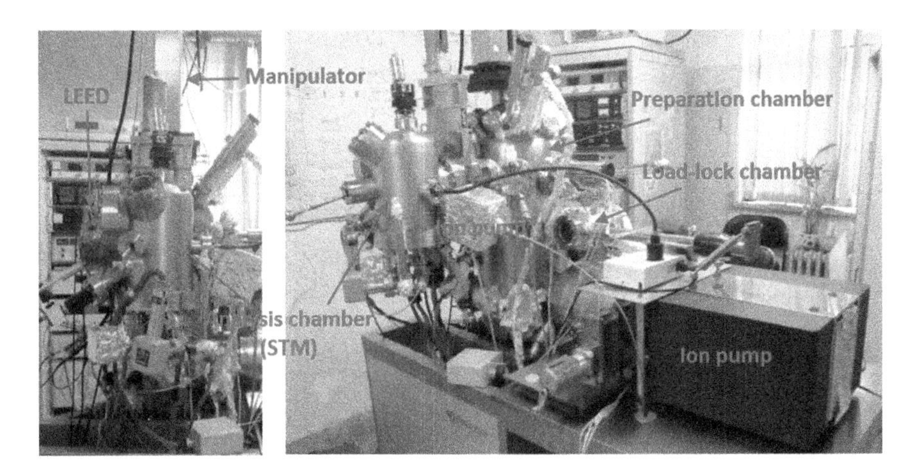

Fig. 1.20 UHV-MBE-VTSTM system

electronic characteristics, which is the chief approach of in situ characterization of sample quality and surface structure. The ancillary facilities include ultra-high vacuum pumps, vibration isolation systems, home-made gas lines, gas station, and so on.

Other characterization techniques, such as scanning transmission electron microscope (STEM), angle-resolved photoemission spectroscopy (ARPES), Raman spectroscopy, and X-ray photoelectron spectroscopy (XPS) were performed in our cooperative groups or public experimental platforms. A home-made sample delivery chamber is used to transfer samples between our chamber and other cooperative UHV systems, in which a UHV environment of 10^{-10} mbar is maintained at all time during sample delivery.

References

1. Novoselov KS et al (2004) Electric field effect in atomically thin carbon films. Science 306:666–669. https://doi.org/10.1126/science.1102896
2. Geim AK, Novoselov KS (2007) The rise of graphene. Nat Mater 6:183–191
3. Neto AHC, Guinea F, Peres NMR, Novoselov KS, Geim AK (2009) The electronic properties of graphene. Rev Mod Phys 81:109–154
4. Zhang YB, Tan YW, Stormer HL, Kim P (2005) Experimental observation of the quantum Hall effect and Berry's phase in graphene. Nature 438:201–204. https://doi.org/10.1038/nature04235
5. Jiang Z, Zhang Y, Tan YW, Stormer HL, Kim P (2007) Quantum Hall effect in graphene. Solid State Commun 143:14–19. https://doi.org/10.1016/j.ssc.2007.02.046
6. Novoselov KS et al (2007) Room-temperature quantum hall effect in graphene. Science 315:1379. https://doi.org/10.1126/science.1137201
7. Balandin AA et al (2008) Superior thermal conductivity of single-layer graphene. Nano Lett 8:902–907. https://doi.org/10.1021/nl0731872
8. Lee C, Wei X, Kysar JW, Hone J (2008) Measurement of the elastic properties and intrinsic strength of monolayer graphene. Science 321:385–388. https://doi.org/10.1126/science.1157996
9. Nair RR et al (2008) Fine structure constant defines visual transparency of graphene. Science 320:1308. https://doi.org/10.1126/science.1156965
10. Schedin F et al (2007) Detection of individual gas molecules adsorbed on graphene. Nat Mater 6:652–655. https://doi.org/10.1038/nmat1967
11. Lin YM et al (2010) 100-GHz transistors from wafer-scale epitaxial graphene. Science 327:662. https://doi.org/10.1126/science.1184289
12. Kim K, Choi J-Y, Kim T, Cho S-H, Chung H-J (2011) A role for graphene in silicon-based semiconductor devices. Nature 479:338–344
13. Lin Y-M et al (2011) Wafer-scale graphene integrated circuit. Science 332:1294–1297. https://doi.org/10.1126/science.1204428
14. Novoselov KS et al (2012) A roadmap for graphene. Nature 490:192–200. https://doi.org/10.1038/nature11458
15. Coraux J et al (2009) Growth of graphene on Ir(111). New J Phys 11:023006. https://doi.org/10.1088/1367-2630/11/2/023006
16. Emtsev KV et al (2009) Towards wafer-size graphene layers by atmospheric pressure graphitization of silicon carbide. Nat Mater 8:203–207. https://doi.org/10.1038/nmat2382
17. Juang Z-Y et al (2009) Synthesis of graphene on silicon carbide substrates at low temperature. Carbon 47:2026–2031. https://doi.org/10.1016/j.carbon.2009.03.051

18. Pan Y et al (2009) Highly ordered, millimeter-scale, continuous, single-crystalline graphene monolayer formed on Ru(0001). Adv Mater 21:2777. https://doi.org/10.1002/adma.200800761
19. Sutter P, Sadowski JT, Sutter E (2009) Graphene on Pt(111): growth and substrate interaction. Phys Rev B 80:245411. https://doi.org/10.1103/PhysRevB.80.245411
20. Bae S et al (2010) Roll-to-roll production of 30-inch graphene films for transparent electrodes. Nat Nanotechnol 5:574–578. https://doi.org/10.1038/nnano.2010.132
21. Murata Y et al (2010) Orientation-dependent work function of graphene on Pd(111). Appl Phys Lett 97:143114. https://doi.org/10.1063/1.3495784
22. Zhu Y et al (2010) Graphene and graphene oxide: synthesis, properties, and applications. Adv Mater 22:3906–3924. https://doi.org/10.1002/adma.201001068
23. Gao M et al (2011) Epitaxial growth and structural property of graphene on Pt(111). Appl Phys Lett 98:033101
24. Geim AK, Grigorieva IV (2013) Van der Waals heterostructures. Nature 499:419–425. https://doi.org/10.1038/nature12385
25. Sipos B et al (2008) From Mott state to superconductivity in 1T-TaS_2. Nat Mater 7:960–965. https://doi.org/10.1038/nmat2318
26. Mak KF, Lee C, Hone J, Shan J, Heinz TF (2010) Atomically thin MoS_2: a new direct-gap semiconductor. Phys Rev Lett 105:136805
27. Yin Z et al (2011) Single-layer MoS_2 phototransistors. ACS Nano 6:74–80
28. Cao T et al (2012) Valley-selective circular dichroism of monolayer molybdenum disulphide. Nat Commun 3:887
29. Mak KF, He K, Shan J, Heinz TF (2012) Control of valley polarization in monolayer MoS_2 by optical helicity. Nat Nanotechnol 7:494–498. https://doi.org/10.1038/nnano.2012.96
30. Wang QH, Kalantar-Zadeh K, Kis A, Coleman JN, Strano MS (2012) Electronics and optoelectronics of two-dimensional transition metal dichalcogenides. Nat Nanotechnol 7:699–712. https://doi.org/10.1038/nnano.2012.193
31. Yang JJ et al (2012) Charge-orbital density wave and superconductivity in the strong spin-orbit coupled $IrTe_2$:Pd. Phys Rev Lett 108:116402. https://doi.org/10.1103/PhysRevLett.108.116402
32. Zeng H, Dai J, Yao W, Xiao D, Cui X (2012) Valley polarization in MoS_2 monolayers by optical pumping. Nat Nanotechnol 7:490–493. https://doi.org/10.1038/nnano.2012.95
33. Butler SZ et al (2013) Progress, challenges, and opportunities in two-dimensional materials beyond graphene. ACS Nano 7:2898–2926. https://doi.org/10.1021/nn400280c
34. Lebegue S, Bjorkman T, Klintenberg M, Nieminen RM, Eriksson O (2013) Two-dimensional materials from data filtering and ab initio calculations. Phys Rev X 3:031002. https://doi.org/10.1103/PhysRevX.3.031002
35. Najmaei S et al (2013) Vapour phase growth and grain boundary structure of molybdenum disulphide atomic layers. Nat Mater 12:754–759. https://doi.org/10.1038/nmat3673
36. Rao CNR, Matte HSSR, Maitra U (2013) Graphene analogues of inorganic layered materials. Angewe Chem Int Ed 52:13162–13185. https://doi.org/10.1002/anie.201301548
37. van der Zande AM et al (2013) Grains and grain boundaries in highly crystalline monolayer molybdenum disulphide. Nat Mater 12:554–561. https://doi.org/10.1038/nmat3633
38. Xu MS, Liang T, Shi MM, Chen HZ (2013) Graphene-like two-dimensional materials. Chem Rev 113:3766–3798. https://doi.org/10.1021/cr300263a
39. Fiori G et al (2014) Electronics based on two-dimensional materials (vol 9, pg 768, 2014). Nat Nanotechnol 9:1063. https://doi.org/10.1038/Nnano.2014.283
40. Koppens FHL et al (2014) Photodetectors based on graphene, other two-dimensional materials and hybrid systems. Nat Nanotechnol 9:780–793. https://doi.org/10.1038/Nnano.2014.215
41. Mak KF, McGill KL, Park J, McEuen PL (2014) The valley Hall effect in MoS_2 transistors. Science 344:1489–1492. https://doi.org/10.1126/science.1250140
42. Xu XD, Yao W, Xiao D, Heinz TF (2014) Spin and pseudospins in layered transition metal dichalcogenides. Nat Phys 10:343–350. https://doi.org/10.1038/Nphys2942

43. Sie EJ et al (2015) Valley-selective optical Stark effect in monolayer WS_2. Nat Mater 14:290–294
44. Helveg S et al (2000) Atomic-scale structure of single-layer MoS_2 nanoclusters. Phys Rev Lett 84:951–954. https://doi.org/10.1103/PhysRevLett.84.951
45. Rao CNR, Nag A (2010) Inorganic analogues of graphene. Eur J Inorg Chem 4244–4250:2010. https://doi.org/10.1002/ejic.201000408
46. Zhu ZY, Cheng YC, Schwingenschlogl U (2011) Giant spin-orbit-induced spin splitting in two-dimensional transition-metal dichalcogenide semiconductors. Phys Rev B 84:153402. https://doi.org/10.1103/PhysRevB.84.153402
47. Liu C-C, Feng W, Yao Y (2011) Quantum spin hall effect in silicene and two-dimensional germanium. Phys Rev Lett 107:076802
48. Chen L et al (2012) Evidence for dirac fermions in a honeycomb lattice based on silicon. Phys Rev Lett 109:056804
49. Feng B et al (2012) Evidence of silicene in honeycomb structures of silicon on Ag(111). Nano Lett 12:3507–3511. https://doi.org/10.1021/nl301047g
50. Fleurence A et al (2012) Experimental evidence for epitaxial silicene on diboride thin films. Phys Rev Lett 108:245501. https://doi.org/10.1103/PhysRevLett.108.245501
51. Gao J, Zhao J (2012) Initial geometries, interaction mechanism and high stability of silicene on Ag(111) surface. Sci Rep 2:861. https://doi.org/10.1038/srep00861
52. Vogt P et al (2012) Silicene: compelling experimental evidence for graphenelike two-dimensional silicon. Phys Rev Lett 108:155501. https://doi.org/10.1103/PhysRevLett.108.155501
53. Meng L et al (2013) Buckled silicene formation on Ir(111). Nano Lett 13:685–690. https://doi.org/10.1021/nl304347w
54. Li L et al (2014) Buckled germanene formation on Pt(111). Adv Mater 26:4820–4824. https://doi.org/10.1002/adma.201400909
55. Splendiani A et al (2010) Emerging photoluminescence in monolayer MoS_2. Nano Lett 10:1271–1275. https://doi.org/10.1021/nl903868w
56. Eda G et al (2011) Photoluminescence from chemically exfoliated MoS_2. Nano Lett 11:5111–5116. https://doi.org/10.1021/nl201874w
57. Radisavljevic B, Radenovic A, Brivio J, Giacometti V, Kis A (2011) Single-layer MoS_2 transistors. Nat Nanotechnol 6:147–150
58. Chhowalla M et al (2013) The chemistry of two-dimensional layered transition metal dichalcogenide nanosheets. Nat Chem 5:263–275. https://doi.org/10.1038/nchem.1589
59. Li LK et al (2014) Black phosphorus field-effect transistors. Nat Nanotechnol 9:372–377. https://doi.org/10.1038/Nnano.2014.35
60. Tao L et al (2015) Silicene field-effect transistors operating at room temperature. Nat Nanotechnol 10:227–231. https://doi.org/10.1038/nnano.2014.325
61. Novoselov K et al (2005) Two-dimensional atomic crystals. Proc Natl Acad Sci U S A 102:10451–10453
62. Coleman JN et al (2011) Two-dimensional nanosheets produced by liquid exfoliation of layered materials. Science 331:568–571
63. Nicolosi V, Chhowalla M, Kanatzidis MG, Strano MS, Coleman JN (2013) Liquid exfoliation of layered materials. Science 340:1226419
64. Li X et al (2011) Large-area graphene single crystals grown by low-pressure chemical vapor deposition of methane on copper. J Am Chem Soc 133:2816–2819. https://doi.org/10.1021/ja109793s
65. Liu K-K et al (2012) Growth of large-area and highly crystalline MoS_2 thin layers on insulating substrates. Nano Lett 12:1538–1544. https://doi.org/10.1021/nl2043612
66. Zhan Y, Liu Z, Najmaei S, Ajayan PM, Lou J (2012) Large-area vapor-phase growth and characterization of MoS_2 atomic layers on a SiO_2 substrate. Small 8:966–971. https://doi.org/10.1002/smll.201102654
67. Lee Y-H et al (2013) Synthesis and transfer of single-layer transition metal disulfides on diverse surfaces. Nano Lett 13:1852–1857. https://doi.org/10.1021/nl400687n

68. Lu X et al (2014) Large-area synthesis of monolayer and few-layer $MoSe_2$ films on SiO_2 substrates. Nano Lett 14:2419–2425. https://doi.org/10.1021/nl5000906
69. Takeda K, Shiraishi K (1994) Theoretical possibility of stage corrugation in Si and Ge analogs of graphite. Phys Rev B 50:14916–14922
70. Cahangirov S, Topsakal M, Aktürk E, Şahin H, Ciraci S (2009) Two- and one-dimensional honeycomb structures of silicon and germanium. Phys Rev Lett 102:236804
71. Aufray B et al (2010) Graphene-like silicon nanoribbons on Ag(110): a possible formation of silicene. Appl Phys Lett 96:183102. https://doi.org/10.1063/1.3419932
72. De Padova P et al (2010) Evidence of graphene-like electronic signature in silicene nanoribbons. Appl Phys Lett 96:261905
73. Tchalala MR et al (2014) Atomic structure of silicene nanoribbons on Ag (110). J Phys Conf Ser 491:012002
74. Lalmi B et al (2010) Epitaxial growth of a silicene sheet. Appl Phys Lett 97:223109
75. Le Lay G, De Padova P, Resta A, Bruhn T, Vogt P (2012) Epitaxial silicene: can it be strongly strained? J Phys D Appl Phys 45:392001
76. Lin C-L et al (2012) Structure of silicene grown on Ag(111). Appl Phys Express 5:045802. https://doi.org/10.1143/apex.5.045802
77. Jamgotchian H et al (2012) Growth of silicene layers on Ag (111): unexpected effect of the substrate temperature. J Phys Condens Matter 24:172001
78. Kuc A, Zibouche N, Heine T (2011) Influence of quantum confinement on the electronic structure of the transition metal sulfide TS_2. Phys Rev B 83:245213. https://doi.org/10.1103/PhysRevB.83.245213
79. Zhang Y et al (2014) Direct observation of the transition from indirect to direct bandgap in atomically thin epitaxial $MoSe_2$. Nat Nanotechnol 9:111–115. https://doi.org/10.1038/Nnano.2013.277
80. Novoselov K (2011) Nobel lecture: graphene: materials in the flatland. Rev Mod Phys 83:837
81. Smith RJ et al (2011) Large-scale exfoliation of inorganic layered compounds in aqueous surfactant solutions. Adv Mater 23:3944–3948
82. Zeng Z et al (2011) Single-Layer Semiconducting Nanosheets: High-Yield Preparation and Device Fabrication. Angew Chem Int Ed 50:11093–11097. https://doi.org/10.1002/anie.201106004
83. Cunningham G et al (2012) Solvent exfoliation of transition metal dichalcogenides: dispersibility of exfoliated nanosheets varies only weakly between compounds. ACS Nano 6:3468–3480
84. Zeng Z et al (2012) An effective method for the fabrication of few-layer-thick inorganic nanosheets. Angew Chem Int Ed 51:9052–9056. https://doi.org/10.1002/anie.201204208
85. Lee YH et al (2012) Synthesis of large-area MoS_2 atomic layers with chemical vapor deposition. Adv Mater 24:2320–2325
86. Shi Y et al (2012) Van der Waals epitaxy of MoS_2 layers using graphene as growth templates. Nano Lett 12:2784–2791
87. Peng Y et al (2001) Hydrothermal synthesis and characterization of single-molecular-layer MoS_2 and $MoSe_2$. Chem Lett 30:772–773
88. Peng Y et al (2001) Hydrothermal synthesis of MoS_2 and its pressure-related crystallization. J Solid State Chem 159:170–173
89. Ramakrishna Matte H et al (2010) MoS_2 and WS_2 analogues of graphene. Angew Chem 122:4153–4156
90. RamakrishnaáMatte H (2011) Graphene analogues of layered metal selenides. Dalton Trans 40:10322–10325
91. Schwierz F (2010) Graphene transistors. Nat Nanotechnol 5:487–496
92. Wu Y et al (2011) High-frequency, scaled graphene transistors on diamond-like carbon. Nature 472:74–78
93. Wu Y et al (2012) State-of-the-art graphene high-frequency electronics. Nano Lett 12:3062–3067
94. Colinge J-P (2004) Multiple-gate soi mosfets. Solid-State Electron 48:897–905

95. Yoon Y, Ganapathi K, Salahuddin S (2011) How good can monolayer MoS_2 transistors be? Nano Lett 11:3768–3773
96. Polman A, Atwater HA (2012) Photonic design principles for ultrahigh-efficiency photovoltaics. Nat Mater 11:174–177
97. Carladous A et al (2002) Light emission from spectral analysis of Au/MoS_2 nanocontacts stimulated by scanning tunneling microscopy. Phys Rev B 66:045401
98. Kirmayer S, Aharon E, Dovgolevsky E, Kalina M, Frey GL (2007) Self-assembled lamellar MoS_2, SnS_2 and SiO_2 semiconducting polymer nanocomposites. Philos Trans R Soc A: Math Phys Eng Sci 365:1489–1508
99. Zeng H et al (2010) "White graphenes": boron nitride nanoribbons via boron nitride nanotube unwrapping. Nano Lett 10:5049–5055
100. Tang Q, Zhou Z (2013) Graphene-analogous low-dimensional materials. Prog Mater Sci 58:1244–1315
101. Kubota Y, Watanabe K, Tsuda O, Taniguchi T (2007) Deep ultraviolet light-emitting hexagonal boron nitride synthesized at atmospheric pressure. Science 317:932–934
102. Yang W et al (2013) Epitaxial growth of single-domain graphene on hexagonal boron nitride. Nat Mater 12:792–797
103. Dean C et al (2010) Boron nitride substrates for high-quality graphene electronics. Nat Nanotechnol 5:722–726
104. Ponomarenko L et al (2011) Tunable metal-insulator transition in double-layer graphene heterostructures. Nat Phys 7:958–961
105. Niu TC, Li A (2015) From two-dimensional materials to heterostructures. Prog Surf Sci 90:21–45. https://doi.org/10.1016/j.progsurf.2014.11.001

Chapter 2
Germanene on Pt(111)

Abstract The tremendous progress in graphene research has motivated us to investigate analogous 2D crystalline systems—new two-dimensionally ordered materials composed of elements other than carbon. Here, this chapter presents the fabrication and structural characterization of germanene on a Pt(111) surface, which is the germanium-based counterpart of graphene. LEED and STM, combined with density functional theory (DFT)-based ab initio calculations reveal the buckled honeycomb-like structure of Pt-supported germanene. Calculated electron localization function (ELF) demonstrates the continuity of 2D germanene on Pt(111).

Keywords Germanene · Germanium · STM · LEED · Honeycomb structure · Pt(111)

2.1 Background

2.1.1 From Silicene to Germanene

The explosive studies on graphene have aroused significant interests towards other graphene-like single-element 2D materials. Graphene is derived from the group IV elements, so it is natural to consider the possibility of graphene analogs based on other group IV elements, such as silicon and germanium. However, silicon and germanium are known to favor sp^3-hybridized, tetrahedral bonds in contrast to planar sp^2 hybridization in graphite. So single layers of silicon and germanium, so-called silicene and germanene, cannot be derived directly from their bulk counterparts. Instead, they have first been fabricated from the computational side. First-principles calculations demonstrated free-standing silicene and germanene favor to adopt a low-buckled honeycomb structure, forming a mixed sp^2–sp^3 like hybridized state [1]. Further theoretical calculations revealed that epitaxial silicene (germanene) on substrates possesses a bandgap of 1.55 meV (23.9 meV) due to the symmetry-breaking and enhanced spin-orbital coupling effects. In particular, a quantum spin Hall effect was predicted in buckled germanene [2], and high-T_c superconductivity was anticipated in doped germanene [3] (where T_c represents the critical temperature for

© Springer Nature Singapore Pte Ltd. 2020

L. Li, *Fabrication and Physical Properties of Novel Two-dimensional Crystal Materials Beyond Graphene: Germanene, Hafnene and PtSe*$_2$, Springer Theses,
https://doi.org/10.1007/978-981-15-1963-5_2

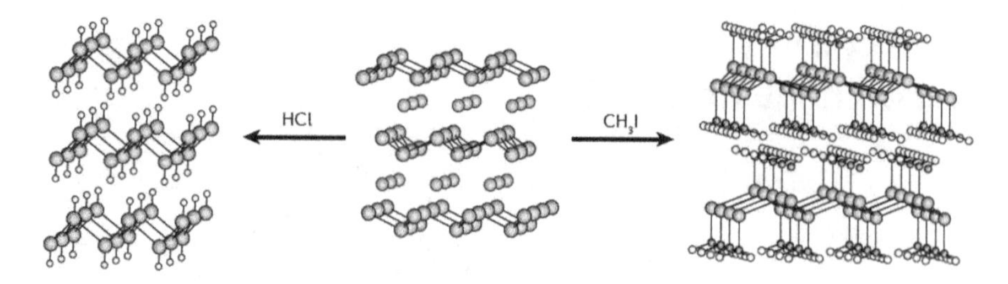

Fig. 2.1 Synthesis of germanane and methylated germanane. Reproduced with permission from Ref. [6], © 2015 ACS; [5], © 2014 Springer Nature

superconductivity). These properties indicate potential applications of silicene and germanene in electronics, spintronics, and photonic devices.

The synthesis of silicene and germanene is the primary step towards the realization of their exotic properties and potential applications. Lack of layered bulk analogs necessitates "bottom-up" growth method, i.e., deposition of silicene and germanene on substrates. The exploration of epitaxial growth of silicon on silver surfaces resulted in the formation of Si nanowires on Ag(100), Si nanoribbons on Ag(110), and silicene on Ag(111). Later, additional studies have reported the fabrication of silicene on Ir(111), ZrB_2 and MoS_2, as discussed in Chap. 1.

2.1.2 *Early Exploration of Germanene Growth*

The preliminary studies of germanene growth started with solution-based exfoliation processes [4]. Chemically treating $CaGe_2$ in HCl created hydrogenated germanene [5], that is, germanane (GeH). Further investigations yielded methylated germanane by treating $CaGe_2$ in CH_3I [6, 7], as depicted in Fig. 2.1. These functionalized germanene sheets showed strong oxidation resistance in ambient conditions. However, these Ge–H and Ge–CH_3 bonds enhance the sp^3 hybridization of Ge atoms, which disables 2D honeycomb structures resulting from sp^2-hybridized orbitals, and more essentially, the linear dispersion relation at the Dirac point. Therefore, these layered materials are not germanene in nature. To the best of our knowledge, the actual fabrication of germanene has not yet been reported, which is our research background and motivation.

2.2 Fabrication and Computation Methods

We grow single-layer germanene sheets on a Pt(111) substrate by surface-assisted molecular beam epitaxial (MBE) [8], as illustrated in Fig. 2.2. The choice of a Pt(111) substrate is based on several advantageous characteristics of Pt(111), such

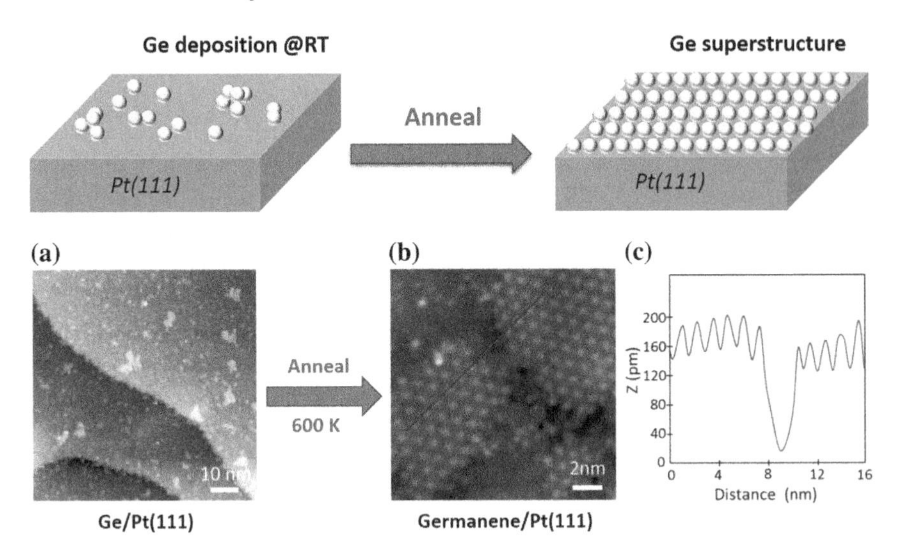

Fig. 2.2 Upper: schematic illustration of physical vapor deposition of germanene on Pt(111). Lower: **a** STM topography of Ge deposited on Pt(111) at room temperature. **b** Ge superstructures formed after annealing at 600 K. **c** The line profile along the blue line in (**b**) showing the thickness of single-layer germanene

as its hexagonal symmetry serving as a growth template and its weaker interfacial interaction with adsorbed 2D honeycomb sheets (e.g. graphene) relative to other metals [9, 10]. Experiments were performed in an ultra-high vacuum (UHV) system with a base pressure of about 2×10^{-10} mbar. The Pt(111) surface was cleaned by several cycles of sputtering and annealing until it yielded distinct Pt(1×1) diffraction spots in the LEED pattern and clean surface terraces in the STM images. The germanium was evaporated onto the Pt(111) substrate kept at room temperature from a high-purity Ge rod mounted in an electron-beam evaporator. The amorphous surface covered by Ge particles and clusters was observed in Fig. 2.2a. The sample was then annealed at a temperature range of 600–750 K for 30 min, which was below 800 K for the sake of excluding the formation of a Ge–Pt surface alloy. STM was then employed to characterize the as-prepared Ge superstructures, as shown in Fig. 2.2b, c. LEED was utilized to identify the superstructures macroscopically.

Our DFT-based first-principle calculations were performed using the Vienna ab initio simulation package (VASP) [11, 12]. The projector augmented wave (PAW) potentials were used to describe the core electrons, and the local density approximation (LDA) was used for exchange and correlation [13]. The periodic slab models included four layers of Pt, one layer of germanium, and a vacuum layer of 15 Å. All atoms were fully relaxed except for the bottom two substrate layers until the net force on every atom was less than 0.01 eV/Å. The energy cutoff of the plane-wave basis sets was 400 eV, and the K-points sampling was $3 \times 3 \times 1$, generated automatically with the origin at the Γ-point.

2.3 Structural Characterizations and Theoretical Calculations

The structure of the germanium layer formed on the Pt(111) surface is characterized macroscopically by LEED, as shown in Fig. 2.3. The six outer symmetric bright spots can be assigned to the pristine Pt(111) substrate, which has a six-fold symmetry. The additional distinct diffraction spots originate from the germanium superstructures. For clarity, we sketched a map of the diffraction spots of the superstructure in reciprocal space (Fig. 2.3b), where the reciprocal vectors of each group of spots are indicated by differently colored arrows. Aside from the (1×1) diffraction spots of the Pt(111) lattice, two symmetrically equivalent domains exist, identified by red and blue spots, respectively. Aiming to understand this LEED pattern more thoroughly, a schematic diagram in real space consistent with the diffraction patterns is provided in Fig. 2.3c. This diagram directly reveals the commensurable relation between the germanium adlayer and the substrate lattice. Germanium forms a superstructure with matrix $\begin{bmatrix} 2 & -3 \\ 3 & 5 \end{bmatrix}$ or the equivalent $\begin{bmatrix} 3 & -2 \\ 2 & 5 \end{bmatrix}$ and the corresponding angles between the close-packed direction of Pt$[1\,\bar{1}\,0]$ and this superstructure can be obtained as 36.6° and 23.4°, respectively. From the data, this superstructure can be easily identified as a $(\sqrt{19} \times \sqrt{19})$ superstructure with respect to the Pt(111) substrate.

In order to characterize the germanium adlayer in detail, we subsequently carried out STM measurements. The large-scale STM image in Fig. 2.4a shows the long-range order of the germanium superstructure formed on the Pt(111) surface. One of the supercells is marked by the blue rhombus. The orientation of the supercell is rotated about 23° relative to Pt$[1\,\bar{1}\,0]$ direction. The corresponding high-resolution

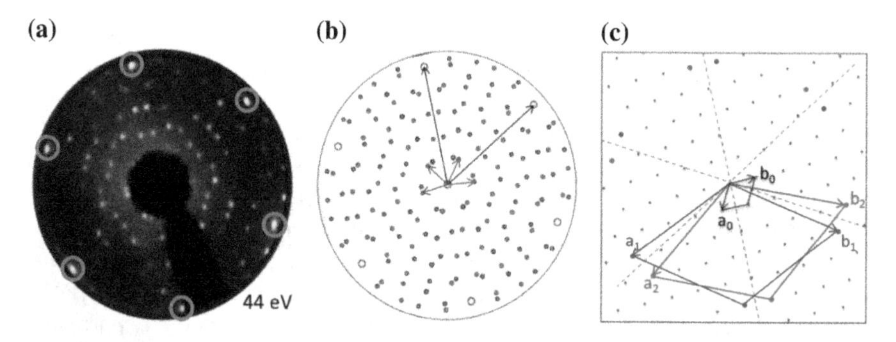

Fig. 2.3 LEED patterns and the corresponding schematic diagram of the germanium superstructure formed on the Pt(111) surface. **a** The six outer spots indicated by circles originate from the six-fold symmetry of the Pt(111) substrate. The additional diffraction spots are ascribed to the germanium adlayer. **b** Schematic representation of the diffraction spots shown in (**a**), where the reciprocal vectors of each group of spots are indicated by black, red, or blue arrows. **c** Schematic diagram of the diffraction spots in real space. These data reveal a $(\sqrt{19} \times \sqrt{19})$ superstructure of the germanium layer [lattice vectors (a_1, b_1) or (a_2, b_2)] with respect to the Pt(111) lattice [lattice vectors (a_0, b_0)]. Reprinted with permission from Ref. [8], © 2014 Wiley

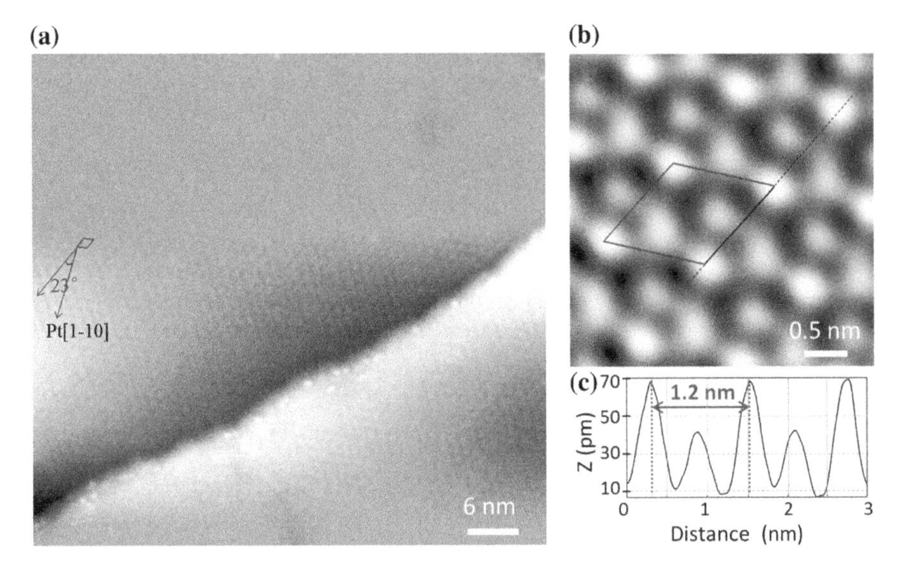

Fig. 2.4 **a** Occupied-state STM image (U = −1.45 V, I = 0.25 nA), showing a ($\sqrt{19} \times \sqrt{19}$) superstructure of the germanium adlayer formed on the Pt(111) surface. The direction of this reconstruction is indicated by the blue arrow. The close-packed direction Pt[1 $\bar{1}$ 0] is indicated by the black arrow. The angle between the blue and black arrows is about 23°. **b** Zoomed-in STM image (U = 1 V, I = 0.05 nA) of the germanium adlayer. **c** Line profile along the dashed line in (**b**), revealing the periodicity of the germanium superstructure (1.2 nm). Reproduced with permission from Ref. [8], © 2014 Wiley

STM image of the germanium adlayer is displayed in Fig. 2.4b, demonstrating two other protrusions with different image contrast inside each supercell. Figure 2.4c shows the line profile along the dashed line of Fig. 2.4b, revealing that the periodicity of the brightest protrusions in the STM image is about 1.2 nm. This distance is equal to the dimension of ($\sqrt{19} \times \sqrt{19}$) superlattice of the Pt(111) surface; the lattice constant of Pt(111) is 0.277 nm ($\sqrt{19} \times 0.277$ nm = 1.21 nm). It is readily apparent that both the orientation and the periodicity of the germanium superstructure detected by STM are in good agreement with the analysis of the LEED pattern, confirming that a ($\sqrt{19} \times \sqrt{19}$) germanium superstructure was formed on the surface. In addition, the profile reveals a corrugation of around 0.6 Å in the germanium adlayer, indicating that different germanium atoms in the adlayer have different apparent heights with respect to the underlying Pt lattice, which we further analyzed using DFT calculations. This corrugation value is close to the buckled height (around 0.64 Å) of free-standing germanene in vacuum [1] and comparable to that of silicene on Ag(111) (0.75 Å) [14]. It implies that the Ge–Pt interaction is similar to the moderate Si–Ag interaction, which facilitates the formation of silicene on Ag(111).

Although Fig. 2.4b is not an STM image at atomic resolution, it is useful for one to construct an atomic model of the germanium adlayer. Note that the lattice constant of low-buckled free-standing germanene is about 3.97–4.02 Å based on the predictions of previous theoretical studies [1, 2, 15]. In that case, the lattice constant of the

(3 × 3) superlattice of the germanene is about 11.91–12.06 Å. This value is close to the periodicity (12 Å) of our observed structure, the ($\sqrt{19} \times \sqrt{19}$) germanium superstructure on Pt(111). Thus, we propose the following model to account for our observations: a single-layer germanene sheet is adsorbed on the Pt(111) surface at a certain rotation angle. As the rotation angle between the close-packed direction of Pt[1 $\bar{1}$ 0] and the ($\sqrt{19} \times \sqrt{19}$) germanium superstructure is determined to be about 23° (Fig. 2.4a), and the angle between the direction of the (1 × 1) lattice of germanene and the direction of its (3 × 3) superlattice is 0°, we can deduce the rotation angle of the high-symmetry direction of the germanium adlayer with respect to the substrate lattice, that is, about 23° between the (1 × 1) lattice of germanene and the Pt[1 $\bar{1}$ 0] direction. In this model, the ($\sqrt{19} \times \sqrt{19}$) supercell of the Pt(111) substrate nearly matches the (3 × 3) superlattice of the germanene sheet.

In order to obtain a detailed structural analysis of the germanium adlayer on Pt(111) and a deep understanding of the germanene/Pt(111) system, we calculated the geometric and electronic structures using DFT-based ab initio calculations, and we performed the corresponding STM simulations using the Tersoff-Hamann approach. Several models (different locations for the germanium atoms with respect to the substrate) were calculated. It turns out that the model with a honeycomb structure (germanene) shown in Fig. 2.5a (top view) is the most stable configuration. The binding energy of this configuration is about −1.39 eV per germanium atom. There are 18 germanium atoms per ($\sqrt{19} \times \sqrt{19}$) unit cell (the yellow rhombus). Different germanium atoms are situated in different chemical environments with respect to the Pt(111) surface. Such differences could account for the overall configuration of the germanene layer. The simulated STM image is shown in Fig. 2.5b, and one unit cell is indicated as a red rhombus. There are four protrusions at the vertices of the rhombus and two at the centers of the two triangular regions of the unit cell. These features are in excellent agreement with the STM observations in Fig. 2.4b, verifying that the model of the (3 × 3) germanene/($\sqrt{19} \times \sqrt{19}$) Pt(111) superstructure fundamentally resembles what we observed in our experiments. The relaxed model and the simulated STM image here provide a clear picture of the germanium arrangements in the

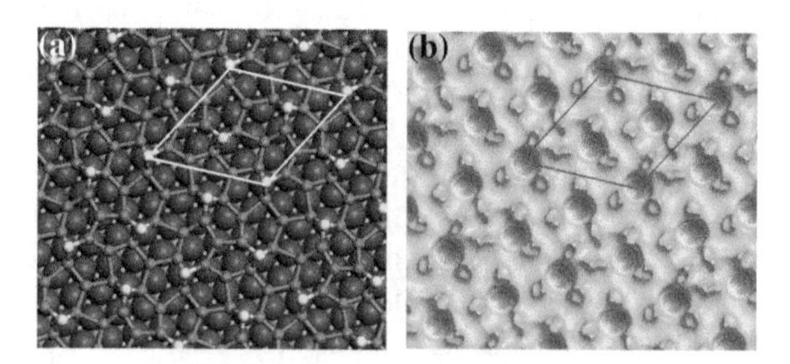

Fig. 2.5 **a** Top view of the relaxed atomic model of the (3 × 3) germanene/($\sqrt{19} \times \sqrt{19}$) Pt(111) configuration. The blue, yellow, and orange spheres represent Pt, protruding Ge, and other Ge atoms, respectively; a unit cell is outlined in yellow. **b** Simulated STM image, showing features identical with the experimental results. The brightness scale is represented by red > white > dark green, and a unit cell is outlined in red. Reproduced with permission from Ref. [8], © 2014 Wiley

adlayer observed in the actual STM images. The germanium atoms at the top sites of the underlying Pt atoms (the yellow spheres in Fig. 2.5a) correspond to the bright protrusions exhibited in the STM image (Fig. 2.4b). Furthermore, in the model, only one of the six germanium atoms in the honeycomb has a higher position, which differs from the theoretical models of a free-standing germanene sheet, wherein half of the germanium atoms have higher positions than the other half [1]. This suggests a substrate-induced buckled conformation of the germanene adlayer.

Although the germanene layer has a buckled structure with an undulation, it is a continuous layer rather than an accumulation of fragments with several germanium atoms, according to the analysis of the electron localization function (ELF). The ELF shows the charge localization between individual atoms, allowing us to appraise the chemical interaction between atoms directly. Figure 2.6a shows the top view of the overall ELF within the germanene layer with an ELF value of 0.5. Here, we see that chemical interaction exists between each pair of germanium atoms, showing the continuity of the germanene layer. This means the germanium atoms are well bonded to each other in the germanene sheet. In order to identify the bonding characteristics within each germanium pair clearly, the ELFs along the cross-section of each Ge–Ge pair (annotated in Fig. 2.6a) are displayed in Fig. 2.6b–f. The ELF values are shown by the color scheme, where red represents the electrons that are highly localized and blue signifies the electrons with almost no localization. It is clearly seen that electrons are localized to a large degree at the Ge–Ge pair containing a top-most Ge atom (the magnitudes of the ELF values are in the range of ~0.75–0.84, as shown in Fig. 2.6b, c, identifying a covalent bond between germanium atoms. The Ge–Ge pair (Fig. 2.6f) with the largest atomic distance has a slightly lower density of electrons localized in its intermediate location (ELF value is about 0.53). In conclusion, the ELFs presented here provide evidence that a covalent interaction exists between the members of each Ge–Ge pair. For comparison, the cross-section of the ELF between the germanium atom at the hexagonally close-packed (hcp) site and its nearest Pt atom is shown in Fig. 2.6g. The distance between such a Ge–Pt pair is the shortest one, and this is the position with the strongest interaction between the germanene layer and the substrate. However, the ELF value is only 0.29 in this case, much smaller than the ELF values of any of the germanium pairs. ELF values of less than 0.5 correspond to an absence of pairing between electrons. Therefore, it can be concluded that the interaction between germanium and the underlying platinum is mainly of an electrostatic origin. This interaction is not strong enough to affect the formation of Ge–Ge bonds and the extension of the germanium sheet. The evidence mentioned above demonstrates that a 2D continuous germanium layer—germanene—was indeed successfully fabricated on Pt(111).

It is worth noting, however, that a Ge–Pt substitutional surface alloy has been claimed previously by Ho et al. [16]. They annealed their sample at 900–1200 K to form a Ge–Pt surface alloy, aiming to obtain a longer catalyst lifetime. In their STM images, only four bright protrusions appear, located at the corners of each ($\sqrt{19} \times \sqrt{19}$) supercell. It is evident that their images show fewer features per unit cell than our high-resolution STM image (Fig. 2.4b). In their model, there is only one Ge atom within each ($\sqrt{19} \times \sqrt{19}$) supercell. This simple model is not suitable for our case,

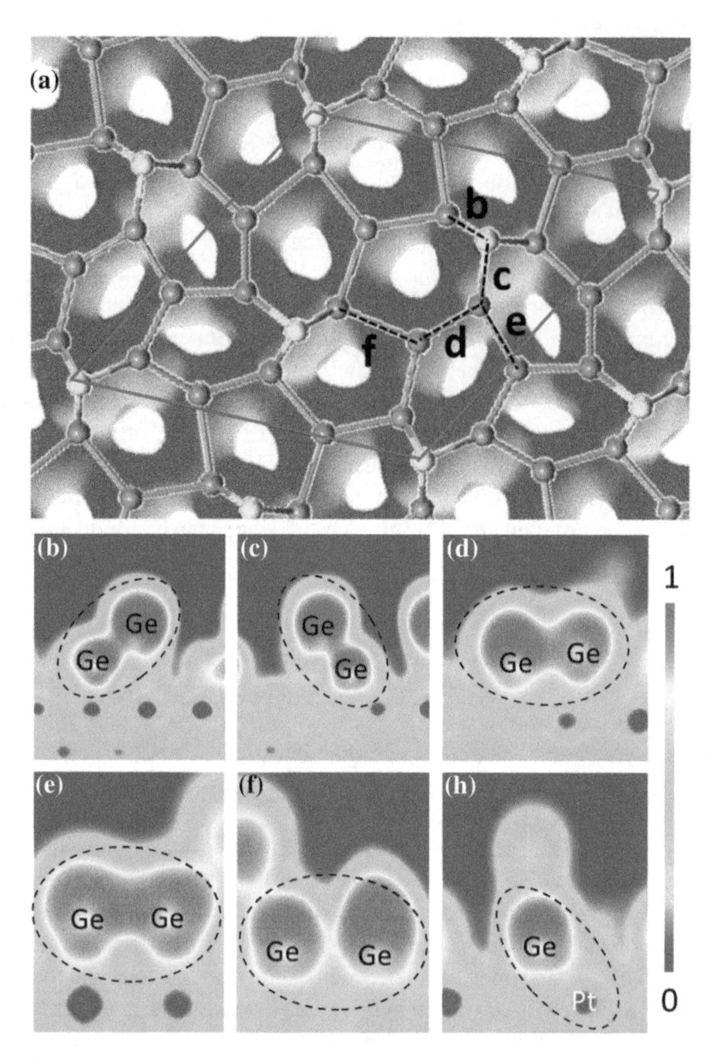

Fig. 2.6 **a** Top view of the overall electron localization function (ELF) of the relaxed model with an ELF value of 0.5, showing continuity of the germanene layer. Yellow and brown spheres represents Ge atoms, with the yellow spheres indicating the protruding Ge atoms. **b–f** The ELFs of the cross-sections between the germanium pairs indicated in (**a**), showing the covalent interaction existing between each pair of germanium atoms. The pairs are depicted by the dashed ovals. **g** The ELF of the cross section between one germanium atom and its nearest Pt neighbor. The ELF value here is in the range of the green-blue region (about 0.29), indicating an electrostatic interaction. The color scale for **b–g** is shown on the right. Reproduced with permission from Ref. [8], © 2014 Wiley

according to our theoretical simulations and experimental observations. We verified that the ($\sqrt{19} \times \sqrt{19}$) superstructure in our case originates not from Ge–Pt chemical contrast but instead from the charge density of state of the buckled germanene layer supported on Pt(111).

2.4 Summary and Outlook

We report for the first time on the fabrication of Ge-based graphene analog-germanene on a Pt(111) substrate.

(1) It features a ($\sqrt{19} \times \sqrt{19}$) superstructure related to the substrate lattice, as demonstrated by LEED and STM. Calculations based on first principles reveal that such a superstructure coincides with the (3×3) superlattice of a buckled germanene sheet.
(2) The calculated electron localization function shows that adjacent germanium atoms directly bind to each other, certifying the formation of a continuous 2D germanene sheet on the Pt(111) surface.
(3) This work provides a method of producing high-quality germanene on solid surfaces so as to explore its physical properties and potential applications in future functional nanodevices.

References

1. Cahangirov S, Topsakal M, Akturk E, Sahin H, Ciraci S (2009) Two- and one-dimensional honeycomb structures of silicon and germanium. Phys Rev Lett 102:236804. https://doi.org/10.1103/PhysRevLett.102.236804
2. Liu C-C, Feng W, Yao Y (2011) Quantum spin hall effect in silicene and two-dimensional germanium. Phys Rev Lett 107:076802
3. Baskaran G (2013) Silicene and germanene as prospective playgrounds for room temperature superconductivity. arXiv:1309.2242
4. Mannix AJ, Kiraly B, Hersam MC, Guisinger NP (2017) Synthesis and chemistry of elemental 2D materials. Nat Rev Chem 1:0014. https://doi.org/10.1038/S41570-016-0014
5. Bianco E et al (2013) Stability and exfoliation of germanane: a germanium graphane analogue. ACS Nano 7:4414–4421
6. Jiang SS et al (2014) Improving the stability and optical properties of germanane via one-step covalent methyl-termination. Nat Commun 5:3389. https://doi.org/10.1038/Ncomms4389
7. Jiang SS, Arguilla MQ, Cultrara ND, Goldberger JE (2015) Covalently-controlled properties by design in group IV graphane analogues. Acc Chem Res 48:144–151. https://doi.org/10.1021/ar500296e
8. Li LF et al (2014) Buckled germanene formation on Pt(111). Adv Mater 26:4820. https://doi.org/10.1002/adma.201400909
9. Gao M et al (2011) Epitaxial growth and structural property of graphene on Pt(111). Appl Phys Lett 98:033101. https://doi.org/10.1063/1.3543624
10. Gao M et al (2010) Tunable interfacial properties of epitaxial graphene on metal substrates. Appl Phys Lett 96:053109. https://doi.org/10.1063/1.3309671

11. Vanderbilt D (1990) Soft self-consistent pseudopotentials in a generalized eigenvalue formalism. Phys Rev B 41:7892
12. Kresse G, Furthmuller J (1996) Efficient iterative schemes for ab initio total-energy calculations using a plane-wave basis set. Phys Rev B 54:11169–11186. https://doi.org/10.1103/PhysRevB.54.11169
13. Perdew JP, Burke K, Ernzerhof M (1996) Generalized gradient approximation made simple. Phys Rev Lett 77:3865–3868. https://doi.org/10.1103/PhysRevLett.77.3865
14. Vogt P et al (2012) Silicene: compelling experimental evidence for graphenelike two-dimensional silicon. Phys Rev Lett 108:155501. https://doi.org/10.1103/PhysRevLett.108.155501
15. Lebegue S, Eriksson O (2009) Electronic structure of two-dimensional crystals from ab initio theory. Phys Rev B 79:115409
16. Ho CS, Banerjee S, Batzill M, Beck DE, Koel BE (2009) Formation and structure of a (root 19 x root 19)R23.4 degrees-Ge/Pt(111) surface alloy. Surf Sci 603:1161–1167. https://doi.org/10.1016/j.susc.2009.01.028

Chapter 3
Hafnene on Ir(111)

Abstract Two-dimensional (2D) honeycomb systems made of elements with d electrons are rare. In this chapter, we report the fabrication of a transition metal (TM) 2D film, namely, hafnium (Hf) monolayer on Ir(111). Experimental characterizations reveal that the Hf layer possesses honeycomb lattice, morphologically identical to graphene. First-principles calculations provide evidence for directional bonding between adjacent Hf atoms, analogous to carbon atoms in graphene. Calculations further suggest that the freestanding Hf honeycomb could be ferromagnetic with magnetic moment $\mu/\mathrm{Hf} = 1.46\ \mu\mathrm{B}$. The realization and investigation of TM honeycomb layers extend the scope of 2D structures and could bring about novel properties for technological applications.

Keywords Hafnium · Honeycomb structure · Transition metal · STM · LEED · Ir(111)

3.1 Background

The novel properties of graphene's honeycomb structure have spurred tremendous interest in investigating other two-dimensional (2D) layered structures beyond graphene. This includes, for example, hexagonal boron nitride [1–3], silicene [4–10], and germanene [11]. Almost exclusively, however, the reported 2D honeycomb materials are made of elements with p-orbital electronic structure [12, 13]. The 2D honeycomb structures made of elements with d electrons are still rare, although there are many more transition metal (TM) elements than the ordinary main group honeycomb elements in the first two rows of the Periodic Table. Generally, the TM elements have rich many-body physical behaviors and coordination chemistry properties; many of them also exist as spin-polarized magnetic ions. Hence, it is highly desirable to fabricate 2D honeycomb structures of the TM elements for drastically enhanced and novel electronic, spintronic, and catalytic activities.

© Springer Nature Singapore Pte Ltd. 2020

L. Li, *Fabrication and Physical Properties of Novel Two-dimensional Crystal Materials Beyond Graphene: Germanene, Hafnene and $PtSe_2$*, Springer Theses,
https://doi.org/10.1007/978-981-15-1963-5_3

3.2 Preparation and STM Study

In the present work, we challenge the conventional wisdom that honeycomb structure can exist only in elements that bear high chemical similarity to carbon. We report that the honeycomb structure of TM elements can also be fabricated with totally uncharted physics and chemistry in particular for Hf on Ir substrate [14]. Low-energy electron diffraction (LEED) and scanning tunneling microscopy (STM) measurements reveal that despite the triangular planar structure of the substrate Hf forms its own honeycomb structure with Hf–Hf nearest neighbor distance remarkably close to that of bulk Hf. Electronic structure calculation further reveals the direct bonding of covalent character between the nearest neighboring Hf atoms, as reflected in the calculated total charge density.

Experiments were carried out in our UHV-MBE-STM system. The Ir(111) single crystal was cleaned by several cycles of Ar^+ ion sputtering followed by annealing at around 1473 K until sharp 1×1 diffraction spots in the LEED pattern and clean surface terraces in the STM images were obtained. Hf atoms, evaporated from the electron-beam evaporator, were deposited at room temperature. The Hf flux was set at 10 nA. The deposition time varied depending on the intended Hf coverage. The sample was subsequently annealed until a well-ordered structure was observed. The STM and LEED techniques were combined to characterize the structural and physical properties of the Hf/Ir system. First-principles calculations were performed using the projector augmented wave (PAW) method as implemented in the Vienna ab initio simulation package (VASP) [15, 16]. The generalized gradient approximation of Perdew, Burke, and Ernzerhof (PBE) was used for the exchange and correlation interaction between electrons [17]. Periodic slabs with four layers of (4×4) Ir(111) were used as the substrate, plus one layer of Hf and a vacuum of 12 Å wide. All atoms were fully relaxed except for the bottom Ir layer until the net force on each atom was less than 0.01 eV/Å. The energy cutoff for the plane-wave basis set is 400 eV, and the k-points sampling is $3 \times 3 \times 1$. To simulate the STM images, we used the Tersoff-Hamann approach.

More specifically, we evaporated Hf atoms onto the clean Ir(111) surface, which, as seen in Fig. 3.1a, formed nanoclusters at room temperature. With follow-up annealing at 673 K, a well-ordered structure was observed in Fig. 3.1b. The structural information of the sample was then characterized by LEED. For comparison, Fig. 3.1c shows a typical LEED pattern of the clean substrate, in which the six spots result from the 6-fold symmetry of the Ir(111). After the deposition of Hf followed by annealing, additional diffraction spots appeared, as indicated by the red arrow in Fig. 3.1d, which signals that a new superstructure of Hf origin has emerged. It has the (2×2) pattern with respect to the Ir(111) substrate. The LEED data thus suggest the formation of a well-ordered network of Hf adlayer with a periodicity twice that of Ir(111).

To gain an understanding of this Hf superstructure in real space, we carried out an STM study. Figure 3.2a shows a typical STM image, which reveals a continuous 2D lattice of the honeycomb structure. The orientation of the honeycomb lattice

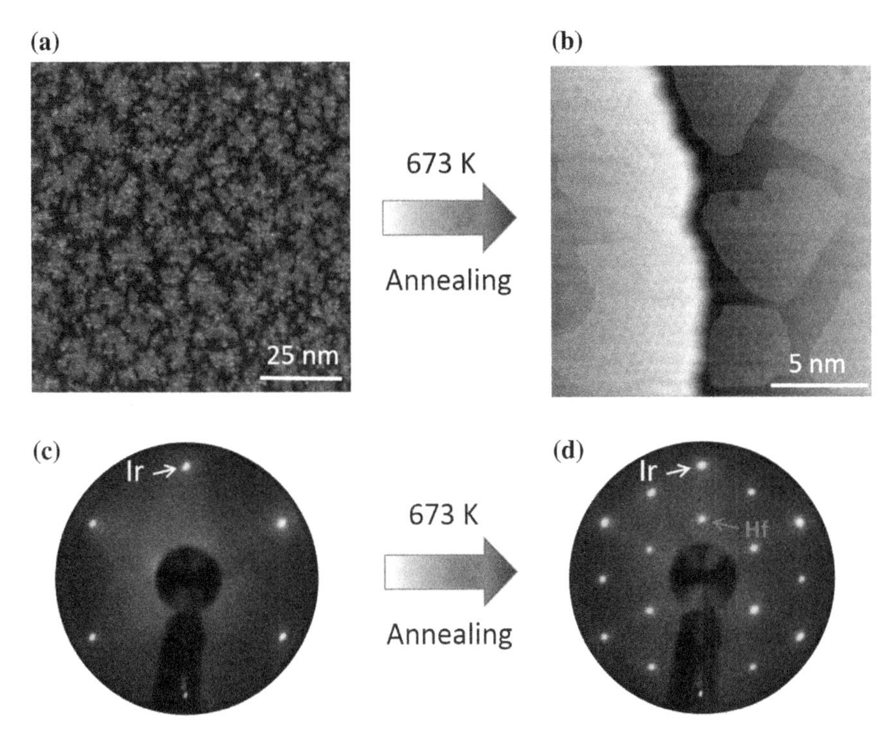

Fig. 3.1 Fabrication of Hafnene on Ir(111). **a** STM morphology of Hf clusters deposited on Ir(111) surface at room temperature (U = −2.0 V and I = 0.2 nA). **b** Hafnene sheets formed by annealing the sample at 673 K (U = −1.2 V and I = 0.5 nA). LEED patterns of **c** the clean Ir(111) substrate and **d** the hafnene sample are obtained at electron beam energies of 68 and 70 eV, respectively. Adapted with permission from Ref. [14], © 2013 ACS

is parallel to the close-packed $[1\,\bar{1}\,0]$ direction of the Ir(111) lattice, in agreement with the LEED results, where the reciprocal vectors of the Ir spots are in line with those of the Hf spots (see Fig. 3.1d). Figure 3.2b is a zoomed-in image in which a perfect honeycomb structure can be clearly resolved. Combining the STM results with the LEED patterns, we conclude that well-ordered Hf honeycomb with a (2×2) superstructure has formed on Ir(111). Figure 3.2c shows a height profile of the honeycomb lattice along the blue line in Fig. 3.2b. It indicates a 5.40 Å periodicity of the honeycomb lattice, which is consistent with the value deduced from the (2×2) spots in the LEED pattern (5.40 Å $\approx 2 \times 2.71$ Å, where 2.71 Å is the lattice constant of the Ir(111)). From this periodicity, we determine the distance between two adjacent Hf atoms in the 2D plane to be $(5.40/\sqrt{3})$ Å = 3.12 Å. This value is very close to the Hf–Hf distance of 3.19 Å on the (0001) facet of bulk Hf. Half of the value, 3.12 Å/2 = 1.56 Å, is also within the range of the proposed covalent radii for Hf, which are between 1.44 and 1.75 Å. In other words, the Hf atoms are the nearest neighbors, rather than disjointed adatoms on Ir(111).

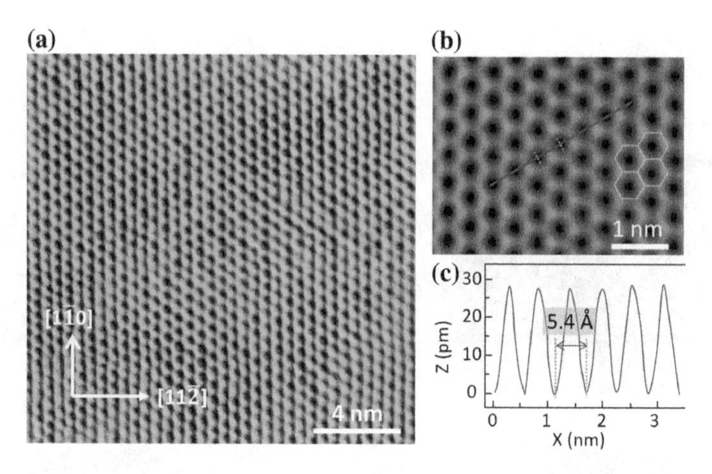

Fig. 3.2 STM images and height profile of Hf layer formed on Ir(111). **a** The topographic image (U = −1.02 V and I = 0.80 nA) shows a well-ordered pattern. The close-packed directions of Ir[1 $\bar{1}$ 0] and [1 1 $\bar{2}$] are indicated by the white arrows. **b** The close-up image (U = −0.70 V and I = 0.16 nA) shows the honeycomb lattice of the Hf adlayer. **c** A height profile taken along the blue line in (**b**), revealing the periodicity of the honeycomb lattice (5.4 Å between holes). Reproduced with permission from Ref. [14], © 2013 ACS

3.3 Theoretical Calculations of Atomic and Electronic Structures

To confirm the experimental observations, we conducted density functional theory (DFT) calculations. For isolated Hf adatoms on Ir(111), we considered three atomic sites, the face-centered cubic (fcc), hexagonal close-packed (hcp), and vertically above an Ir substrate atom (atop), as the building units of the honeycomb lattice. Figure 3.3a demonstrates that Hf binding to the atop site is the weakest; relative to the atop site, the binding energies of the hcp and fcc sites are 1.54 and 1.28 eV/Hf, respectively. Next, we mixed the hcp, fcc, and atop sites to obtain three honeycomb structures, as found in Fig. 3.3b. Not surprisingly, the fcc/hcp mixing is 0.45 and 0.36 eV/Hf lower in energy than the atop/fcc and atop/hcp mixings, respectively.

Figure 3.4a illustrates a top view of the atomic structure for the fcc/hcp mixing on Ir(111), whereas Fig. 3.4b shows the corresponding simulated STM image for bias = −1.5 V. The honeycomb structure is clearly seen in the STM image as marked by the green hexagon. Figure 3.4c displays the corresponding experimental STM image. The overall features are in remarkable agreement with the simulated results in Fig. 3.4b.

A key issue here is whether Hf forms its own honeycomb lattice. While both the experimental STM images in Figs. 3.2b and 3.4c and the measured Hf–Hf distance strongly suggest so, more evidence comes from the calculated charge distribution in the Hf layer. Figure 3.5 depicts the total charge density in the Hf honeycomb plane. It reveals the formation of direct Hf–Hf bonds and thus explains why the Hf

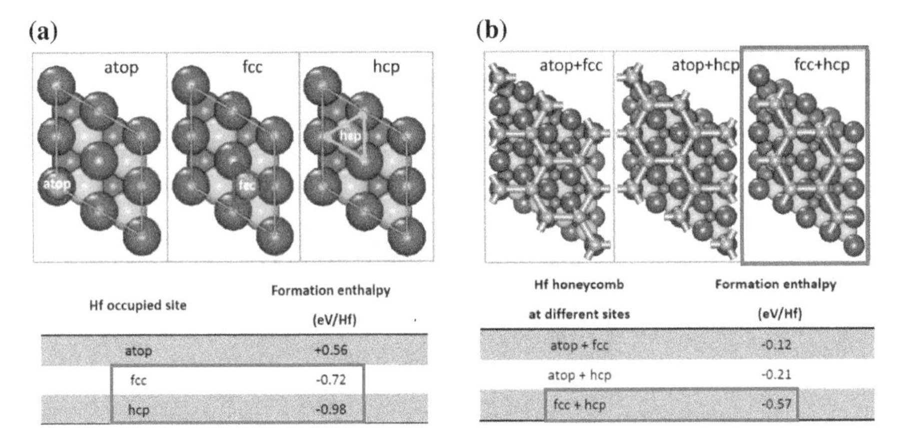

Hf occupied site	Formation enthalpy (eV/Hf)
atop	+0.56
fcc	-0.72
hcp	-0.98

Hf honeycomb at different sites	Formation enthalpy (eV/Hf)
atop + fcc	-0.12
atop + hcp	-0.21
fcc + hcp	-0.57

Fig. 3.3 Theoretical calculations for formation enthalpy of **a** single Hf atoms and **b** a layer of Hf honeycomb on Ir(111). Reproduced with permission from Ref. [14], © 2013 ACS

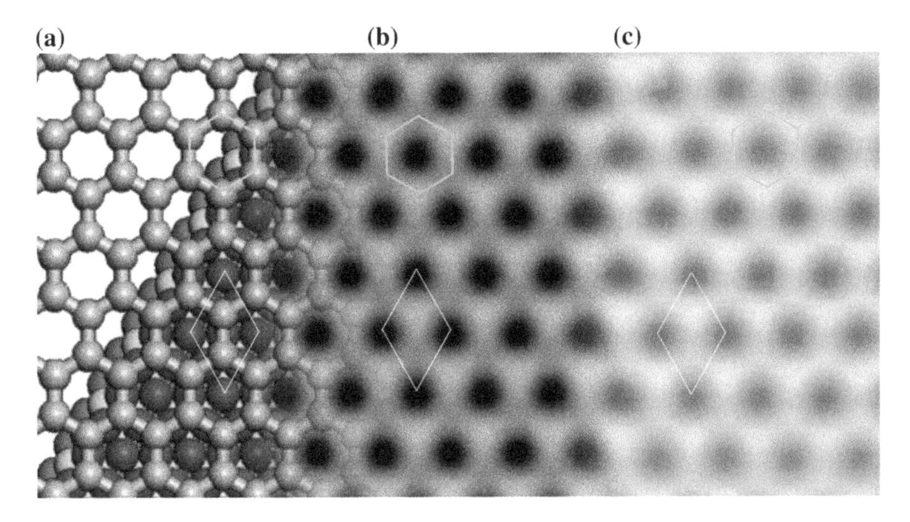

Fig. 3.4 Atomic configuration of Hf honeycomb lattice on Ir(111). **a** Top view of the calculated atomic structure. Half of the Hf atoms are on the fcc sites (vertically above the red Ir balls), while the other half are on the hcp sites (vertically above the cyan Ir balls). The white rhombus denotes the Ir(111) − (2 × 2) unit cell. **b** DFT-simulated STM image (−1.5 V). The Hf honeycomb is highlighted by the green hexagon. **c** Atomically resolved experimental STM image (−1.5 V, 0.1 nA) showing features identical to those in (**b**). Reproduced with permission from Ref. [14], © 2013 ACS

does not merely follow the Ir triangular lattice. Another issue is whether Hf forms the honeycomb with Ir (in other words, forming surface alloys). Our calculations, as exhibited in Fig. 3.6, reveal that a honeycomb made of only Hf is energetically more stable than a honeycomb with Hf–Ir mixing. Therefore, both experiment and theory suggest the existence of a continuous single-layer Hf film with honeycomb

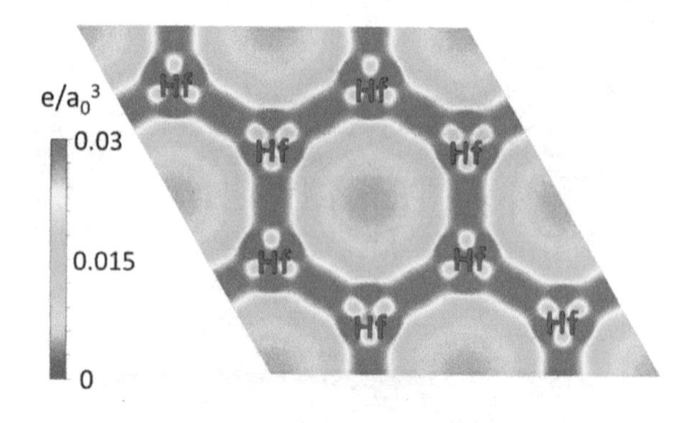

Fig. 3.5 The 2D charge density in the Hf plane on Ir(111) substrate. Reprinted with permission from Ref. [14], © 2013 ACS

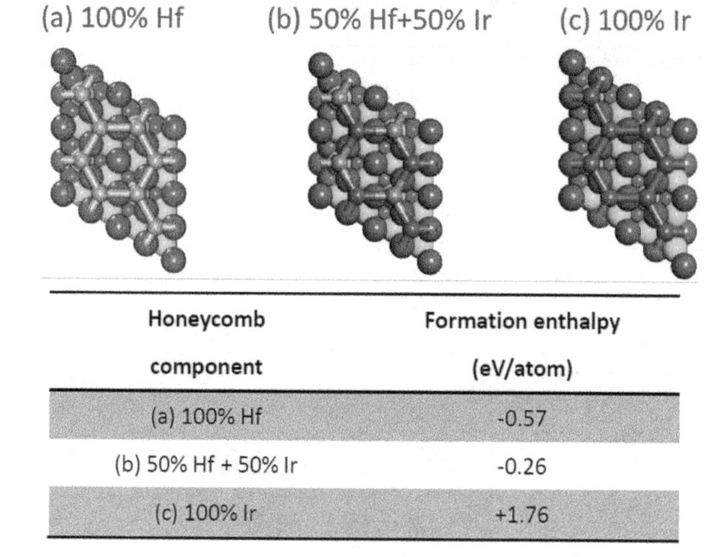

Honeycomb component	Formation enthalpy (eV/atom)
(a) 100% Hf	-0.57
(b) 50% Hf + 50% Ir	-0.26
(c) 100% Ir	+1.76

Fig. 3.6 Atomic models and formation enthalpy of honeycomb structures with different Hf–Ir mixing. Reproduced with permission from Ref. [14], © 2013 ACS

structure, supported on the Ir(111) substrate. The name of this material may be coined as "hafnene", as a particular case of possible metalenes.

In light of the fact that freestanding honeycomb structures usually hold essential properties for technological applications, we have calculated the band structures of a freestanding Hf honeycomb. Figure 3.7 shows the band structures for (a) spin-up and (b) spin-down states. They reveal that the freestanding Hf honeycomb is metallic and strongly spin-polarized (and hence ferromagnetic), with a magnetic moment of 1.46 μB/Hf. The density of states (DOS) plots show the presence of a large number of d states near the Fermi level, which could be utilized for catalytic reactions [18]. Dirac cones also exist in this system. Because of the localization of the wave functions

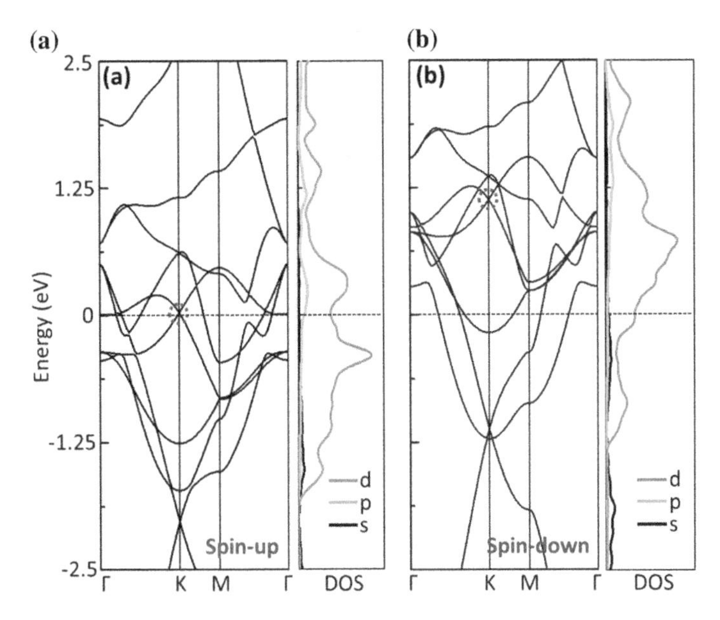

Fig. 3.7 Band structures of free-standing Hf honeycomb. **a** Spin-up and **b** spin-down states along with their partial densities of states. Dotted cycles denote the positions of the Dirac cones near the Fermi level (set at E = 0). Reproduced with permission from Ref. [14], © 2013 ACS

in the 2D plane, which increases exchange coupling between electrons, the Dirac cone for the spin-down states is about 1 eV higher in energy than that for the spin-up states.

3.4 Bilayer Hafnene and Hafnene Grown on Other Supports

We observed the formation of a second Hf honeycomb layer on top of the first one upon increasing the Hf coverage, as shown in Fig. 3.8b. This could be important, as it suggests that the existence of the Hf honeycomb structure does not necessarily depend on the binding to the Ir substrate. Thus there comes a question that whether the single-layer Hf honeycomb structure could readily be separated from the substrate. It has not been possible so far, and even if separated, the freestanding film may not be stable. However, there may be other ways to isolate the Hf honeycomb structure from the substrate without strongly perturbing its physical properties or its physical integrity, for example, by intercalation [19–21].

What is more, it is necessary to note that the observation of Hf honeycomb on Ir(111) is not an isolated case. A similar honeycomb structure has also been fabricated on Rh(111) substrate, as revealed in Fig. 3.9 by LEED and STM.

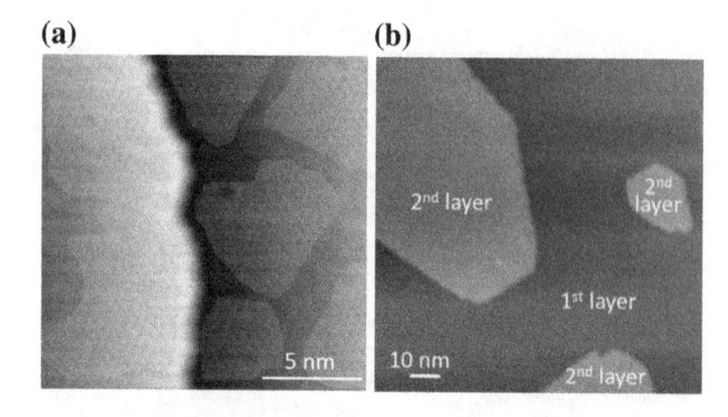

Fig. 3.8 **a** Sub-monolayer and **b** bilayer hafnene. Adapted with permission from Ref. [14], © 2013 ACS

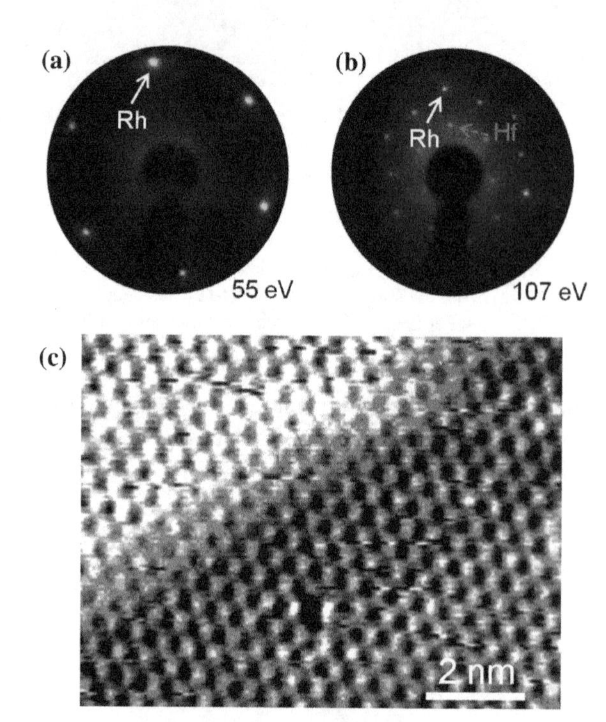

Fig. 3.9 LEED patterns of the Rh substrate **a** before and **b** after hafnene deposition. **c** The STM image of hafnene lattice on Rh(111). Reproduced with permission from Ref. [14], © 2013 ACS

3.5 Summary and Outlook

We report the first example of transition metal honeycomb structures-hafnene.

(1) The LEED and STM studies reveal highly ordered Hf honeycomb lattices on Ir(111) substrates.
(2) Interesting ferromagnetism, high density of d states at the Fermi level, and large spin-splitting of the Dirac cones are theoretically predicted for a freestanding Hf honeycomb structure.

(3) Given the rich electronic, magnetic, and catalytic properties of TM elements in general, and the insufficiency of knowledge of their 2D structures, further realization and investigation of TM single layers are highly desirable. As the TM honeycomb structures provide a new platform to investigate hitherto unknown quantum phenomena and electronic behaviors in 2D systems, we expect them to have broad application potential in nanotechnology and related areas.

References

1. Corso M et al (2004) Boron nitride nanomesh. Science 303:217–220
2. Dean C et al (2010) Boron nitride substrates for high-quality graphene electronics. Nat Nanotechnol 5:722–726
3. Britnell L et al (2012) Field-effect tunneling transistor based on vertical graphene heterostructures. Science 335:947–950
4. Cahangirov S, Topsakal M, Akturk E, Sahin H, Ciraci S (2009) Two- and one-dimensional honeycomb structures of silicon and germanium. Phys Rev Lett 102:236804. https://doi.org/10.1103/PhysRevLett.102.236804
5. Liu C-C, Feng W, Yao Y (2011) Quantum spin hall effect in silicene and two-dimensional germanium. Phys Rev Lett 107:076802
6. Chen L et al (2012) Evidence for dirac fermions in a honeycomb lattice based on silicon. Phys Rev Lett 109:056804
7. Feng B et al (2012) Evidence of silicene in honeycomb structures of silicon on Ag(111). Nano Lett 12:3507–3511. https://doi.org/10.1021/nl301047g
8. Fleurence A et al (2012) Experimental evidence for epitaxial silicene on diboride thin films. Phys Rev Lett 108:245501. https://doi.org/10.1103/PhysRevLett.108.245501
9. Vogt P et al (2012) Silicene: compelling experimental evidence for graphenelike two-dimensional silicon. Phys Rev Lett 108:155501. https://doi.org/10.1103/PhysRevLett.108.155501
10. Tao L et al (2015) Silicene field-effect transistors operating at room temperature. Nat Nanotechnol 10:227–231. https://doi.org/10.1038/nnano.2014.325
11. Li L et al (2014) Buckled germanene formation on Pt(111). Adv Mater 26:4820–4824. https://doi.org/10.1002/adma.201400909
12. Mannix AJ, Kiraly B, Hersam MC, Guisinger NP (2017) Synthesis and chemistry of elemental 2D materials. Nat Rev Chem 1:0014. https://doi.org/10.1038/s41570-016-0014
13. Molle A et al (2017) Buckled two-dimensional Xene sheets. Nat Mater 16:163–169. https://doi.org/10.1038/nmat4802
14. Li LF et al (2013) Two-dimensional transition metal honeycomb realized: Hf on Ir(111). Nano Lett 13:4671–4674. https://doi.org/10.1021/nl4019287
15. Vanderbilt D (1990) Soft self-consistent pseudopotentials in a generalized eigenvalue formalism. Phys Rev B 41:7892
16. Kresse G, Furthmuller J (1996) Efficient iterative schemes for ab initio total-energy calculations using a plane-wave basis set. Phys Rev B 54:11169–11186. https://doi.org/10.1103/PhysRevB.54.11169
17. Perdew JP, Burke K, Ernzerhof M (1996) Generalized gradient approximation made simple. Phys Rev Lett 77:3865–3868. https://doi.org/10.1103/PhysRevLett.77.3865
18. Hammer B, Nørskov JK (2000) Theoretical surface science and catalysis—calculations and concepts. Adv Catal 45:71–129

19. Huang L et al (2011) Intercalation of metal islands and films at the interface of epitaxially grown graphene and Ru (0001) surfaces. Appl Phys Lett 99:163107
20. Meng L et al (2012) Silicon intercalation at the interface of graphene and Ir (111). Appl Phys Lett 100:083101
21. Li L, Wang Y, Meng L, Wu R-T, Gao H-J (2013) Hafnium intercalation between epitaxial graphene and Ir (111) substrate. Appl Phys Lett 102:093106

Chapter 4
Monolayer $PtSe_2$

Abstract Single-layer transition-metal dichalcogenides (TMDs) attract considerable attention due to their interesting physical properties and potential applications. Here, we demonstrate the epitaxial growth of high-quality monolayer platinum diselenide ($PtSe_2$), a new member of the layered TMDs family, by a single step of direct selenization of a Pt(111) substrate. A combination of atomic-resolution experimental characterizations and first-principles theoretic calculations reveals the atomic structure of the monolayer $PtSe_2$/Pt(111). Angle-resolved photoemission spectroscopy measurements confirm for the first time the semiconducting electronic structure of monolayer $PtSe_2$ (in contrast to its semimetallic bulk counterpart). The photocatalytic activity of monolayer $PtSe_2$ film is evaluated by a methylene-blue photodegradation experiment, demonstrating its practical application as a promising photocatalyst. Moreover, circular polarization calculations predict that monolayer $PtSe_2$ also has potential applications in valleytronics.

Keywords $PtSe_2$ · Transition metal dichalcogenides · Atomic structure · Electronic property · Photocatalysis · Valleytronics

4.1 Background

As discussed in Chap. 1, layered transition-metal dichalcogenides (TMDs) with the general formula MX_2, where M represents a transition metal from groups 4–10 and X is a chalcogen (S, Se, or Te), received significant attention in the last dozen years due to their intriguing physical properties for both fundamental research and potential applications in electronics, optoelectronics, spintronics, catalysis, and so on. Depending on the coordination environment and oxidization state of the transition metal, layered TMDs can be metals, semiconductors, and insulators and thus show various physical properties. Recent investigations of MX_2 have resulted in discoveries of dramatically different electronic structures at the monolayer limit compared to the bulk materials due to quantum confinement effects. For example, while pushing from bulk to monolayer, MoS_2 and $MoSe_2$ show an indirect-to-direct bandgap transition [1–4]. With these exciting findings, experimental research efforts so far have been mainly focused on prototypical semiconducting MX_2 with group VIB transition

© Springer Nature Singapore Pte Ltd. 2020

L. Li, *Fabrication and Physical Properties of Novel Two-dimensional Crystal Materials Beyond Graphene: Germanene, Hafnene and $PtSe_2$*, Springer Theses,
https://doi.org/10.1007/978-981-15-1963-5_4

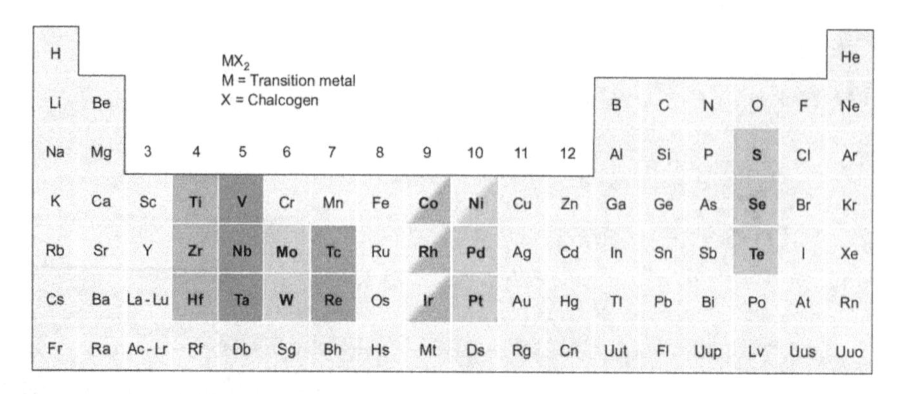

Fig. 4.1 Summary of about 40 different layered MX_2 compounds, which are highlighted. Partial highlights for Co, Rh, Ir and Ni indicate that only some of the dichalcogenides form layered structures. Reprinted with permission from Ref. [6], © 2013 Springer Nature

metals (M = Mo, W). Note, however, that about 40 different MX_2 compounds can form stable, 2D single-layer TMDs structures [5, 6], as summarized in Fig. 4.1. In the large family of layered TMDs, many other promising single-layer TMDs and related quantum defined properties remain to be explored experimentally. For example, $IrTe_2$ and 1T-TaS_2 exhibit novel low-temperature phenomena including superconductivity and charge density wave [7, 8]; bulk ReS_2 shows monolayer behavior due to electronic and vibrational decoupling [9]. These interesting properties motivate considerable interest in exploring other promising TMDs materials, such as group 10 TMDs, which have rarely been reported.

Reliable preparation of ultra-thin 2D TMDs is the essential step for exploring their properties and applications. Among various production methods, chemical vapor deposition (CVD) and CVD analogs are the most important approaches of bottom-up synthesis. As described in Sect. 1.3.4.2, MX_2 growth using CVD-related methods usually involves two components containing M and X as precursors, which increases experimental steps and complexities.

4.2 Growth and Atomic Structure

In this chapter, we report epitaxial growth of monolayer $PtSe_2$ - a heretofore-unexplored member of the single-layer TMDs family—on a Pt substrate by direct "selenization" [10], an analog of direct oxidation. In contrast to conventional fabrication methods of MX_2 by exfoliation or chemical vapor deposition, the present route toward a monolayer dichalcogenide is very straightforward: only one element, Se, is deposited on a Pt(111) substrate, and then the sample is annealed to ~200 °C to obtain epitaxial $PtSe_2$ films, as illustrated in Fig. 4.2a.

The growth of $PtSe_2$ thin films was monitored by in situ X-ray photoelectron spectroscopy (XPS). Figure 4.2b shows the XPS spectra of the Se 3d core level during

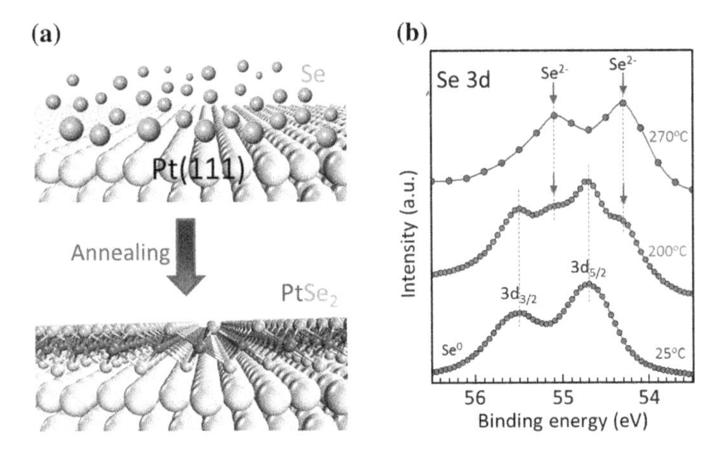

Fig. 4.2 **a** Schematic illustration of synthesizing $PtSe_2$ monolayer by a single step of direct selenization of a Pt(111) substrate. The Pt spheres with different colors and sizes are used just to differentiate the Pt atoms in the Pt(111) substrate and in the $PtSe_2$ sublayer. **b** XPS measurements for the binding energies of Se during $PtSe_2$ growth demonstrating the formation of $PtSe_2$ at 270 °C. The blue arrows indicate the peak positions (55.19 and 54.39 eV) corresponding to the binding energy of Se^{2-}. The Se^0 peaks (at 55.68 and 54.80 eV) are dominant at 25 °C, whereas at 200 °C the peaks in the curve indicate the coexistence of Se^0 and Se^{2-}. Reprinted with permission from Ref. [10], © 2015 ACS

$PtSe_2$ growth. When Se-deposited Pt(111) substrate is annealed to 200 °C, two new peaks appear at binding energies of 55.19 and 54.39 eV (labeled by the blue arrows), which can be explained by a change in the chemical state of Se from Se^0 to Se^{2-}, corresponding to the selenization process of the sample. Further annealing of the sample to 270 °C results in the disappearance of Se^0 peaks (at 55.68 and 54.80 eV) and the dominance of Se^{2-} peaks, indicating full crystallization and complete formation of $PtSe_2$ films.

To obtain the structural information on the as-grown epitaxial films, we observed the samples by LEED. Figure 4.3a shows a LEED pattern. Hexagonal diffraction spots from $PtSe_2$ (red circles) are observed to have the same orientation as those from the Pt(111) substrate (blue circles), suggesting a rotational-domain-free growth. A (3 × 3) diffraction pattern of the epitaxial $PtSe_2$ film is clearly identified, which corresponds to a well-defined moiré superstructure arising from the lattice mismatch between the $PtSe_2$ film and Pt(111) substrate. Furthermore, identical LEED patterns were observed on the entire sample surface (4 mm × 4 mm in size), indicating the growth of a large-area, homogeneous, and high-quality film.

To investigate the atomic structure of the $PtSe_2$ film, we performed STM studies. Figure 4.3b is a large-scale STM image with a well-ordered moiré pattern of $PtSe_2$ thin film on Pt(111). The periodicity of this moiré pattern is about 11.1 Å, four times the lattice constant of Pt(111). Figure 4.3c shows an atomic-resolution image of the area indicated by the white square in Fig. 4.3b, revealing hexagonally arranged protrusions with an average lattice constant of $a_1 = 3.7$ Å, which agrees perfectly

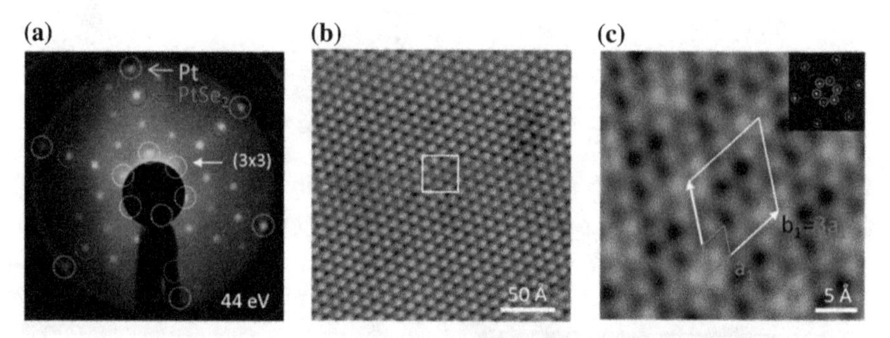

Fig. 4.3 **a** LEED pattern of $PtSe_2$ films formed on the Pt substrate. The blue, red, white circles indicate the diffraction spots from the Pt(111) lattice, $PtSe_2$ thin film, and (3 × 3) superstructure with respect to $PtSe_2$, respectively. **b** Large-scale STM image (U = −1.9 V, I = 0.12 nA) shows the moiré pattern of $PtSe_2$ thin film on Pt(111). The white rectangle marks the size of the close-up image in (**c**). **c** Atomic-resolution STM image (U = −1.0 V, I = 0.12 nA) of single-layer $PtSe_2$ showing the hexagonal lattice of Se atoms in the topmost sublayer of the $PtSe_2$ sandwich-type structure. A (3 × 3) moiré superstructure is visible. The red and white rhombi denote the unit cell of the $PtSe_2$ lattice and (3 × 3) superlattice, respectively. The inset displays the FFT pattern corresponding to $PtSe_2$ and the superstructure. Adapted with permission from Ref. [10], © 2015 ACS

with the interatomic spacing of Se atoms in the (0001) basal plane of bulk $PtSe_2$. Therefore, we interpret the hexagonal protrusions in Fig. 4.3c to be the Se atoms in the topmost Se plane of a $PtSe_2$ film. A regular (3 × 3) moiré superstructure with respect to the $PtSe_2$ lattice is then established, with a periodicity of $b_1 = 3a_1 \cong 11.1$ Å (labeled by the white rhombus). The orientation of the moiré pattern is aligned with that of the $PtSe_2$ lattice. This is in agreement with the LEED observation (Fig. 4.3a), where the diffraction spots of the (3 × 3) superlattice are in line with those of the $PtSe_2$ lattice. Based on LEED and STM measurements, the moiré pattern can be explained as the (3 × 3) $PtSe_2$ supercells located on the (4 × 4) Pt(111) atoms.

To gain further insight into interfacial features of the $PtSe_2$/Pt(111) sample, we performed a cross-section high-angle annular-dark-field (HAADF) STEM study. A Z-contrast image of the $PtSe_2$/Pt(111) interface is shown in Fig. 4.4a. One bright layer combined with two dark layers observed on the topmost surface suggests a Se–Pt–Se sandwich configuration (as indicated by a model diagram superimposed in Fig. 4.4a). The atomically resolved bulk Pt substrate lattice with an experimentally measured interlayer spacing of 2.28 Å served as a reference for calibrating other spacing measurements. After the calibration, the spacing between the Se sublayers in the Se–Pt–Se sandwich is found to be 2.53 Å, which is in agreement with the calculated value in single-layer $PtSe_2$, as displayed in Fig. 4.4b. These atomic-scale cross-section data obtained by STEM further verify that the fabricated structure is indeed a single-layer $PtSe_2$ film on the Pt(111) substrate.

The combination of LEED, STM, and STEM studies indicates a (3 × 3) single-layer $PtSe_2$ on a (4 × 4) Pt(111) structure. We then carried out DFT calculations based on this structure. The simulated STM image is shown in Fig. 4.4c, in which the overall features of the experimental STM image (Fig. 4.3c) are well reproduced.

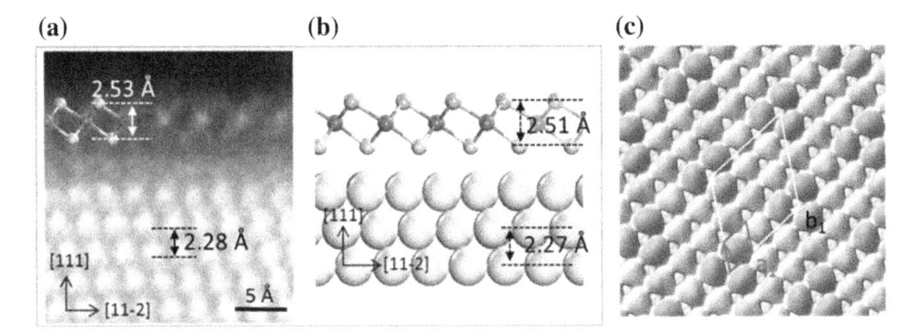

Fig. 4.4 **a** Atomic-resolution STEM Z-contrast image of the $PtSe_2$/Pt(111) interface along the [110] zone axis. A $PtSe_2$ single layer over the Pt(111) substrate is resolved at atomic scale with a model diagram overlaid for clarity. **b** The relaxed model. The blue and orange spheres represent Pt atoms and Se atoms, respectively. **c** Simulated STM image (U = −1.0 V) based on the calculated atomic structure in (**b**) is consistent with the experimental observation in Fig. 4.3c. Adapted with permission from Ref. [10], © 2015 ACS

Only the hexagonally arranged Se atoms in the topmost sublayer of monolayer $PtSe_2$ are imaged. The remarkable agreement between the STM simulation and experimental STM observation strongly supports our conclusions and thus demonstrates the successful growth of a highly crystalline $PtSe_2$ monolayer.

4.3 Electronic Structure

Having grown highly crystalline single-layer $PtSe_2$, we investigated its electronic energy band structure by ARPES. Figure 4.5a shows ARPES data measured along the high symmetry direction K-Γ-M-K in the hexagonal Brillouin zone at a photon energy of 21.2 eV. Data taken at other photon energies show the same dispersion, confirming the 2D character of the monolayer $PtSe_2$. Second-derivative spectra of raw experimental band structures (Fig. 4.5a) are depicted in Fig. 4.5b to enhance the visibility of the bands. Here, the top of the valence band is observed to be at −1.2 eV at the Γ point and the conduction band is above the Fermi level, indicating that monolayer $PtSe_2$ is a semiconductor. It is quite different from the bulk $PtSe_2$, which-according to calculations-is a semimetal. A direct comparison between the ARPES spectrum (Fig. 4.5b) and the calculated band structure (green dotted lines in Fig. 4.5b) shows excellent quantitative agreement. Combining the ARPES spectra with DFT calculations, we confirm that we have synthesized a single-layer $PtSe_2$ and that the epitaxial $PtSe_2$ essentially has the same electronic properties as the free-standing single-layer $PtSe_2$. For the first time, the band structure of monolayer $PtSe_2$ films has been determined experimentally.

The semimetal-to-semiconductor transition was revealed by DFT-LDA calculations. As shown in Fig. 4.5c, the band structure and density of state (DOS) suggest

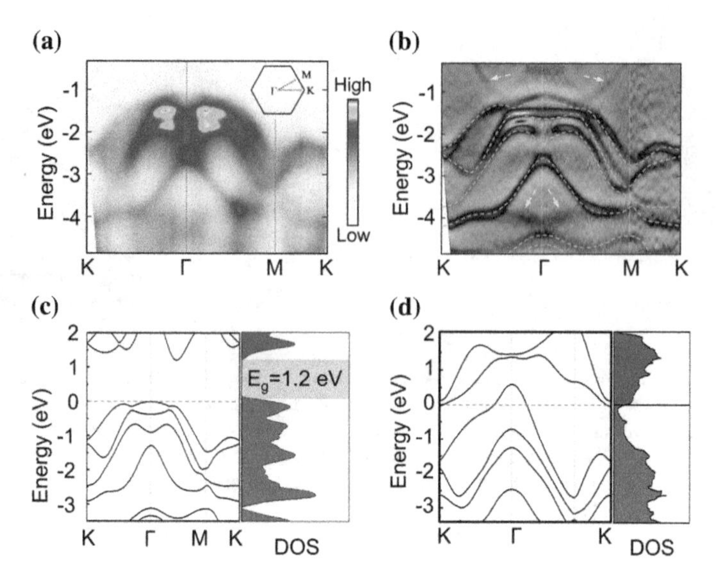

Fig. 4.5 ARPES spectra and valence bands of single-layer $PtSe_2$. **a** ARPES spectra obtained on the monolayer $PtSe_2$ on Pt(111). The high symmetry directions are shown in the inset. **b** Second-derivative spectra of the raw ARPES data in (**a**). The calculated valence bands, superimposed as green dashed lines, are in excellent agreement with the experimental data. The bands marked by white arrows are from the Pt substrate. **c, d** Theoretically calculated band structure and density of states of monolayer $PtSe_2$ and bulk $PtSe_2$, respectively. Reproduced with permission from Ref. [10], © 2015 ACS

single-layer $PtSe_2$ is a semiconductor with an energy gap of 1.20 eV (2.10 eV bandgap is predicted from GW calculation [11]). As a comparison, the band structure and DOS of bulk $PtSe_2$ are plotted in Fig. 4.5d, which confirm that bulk $PtSe_2$ is semimetallic [12, 13]. Actually, bilayer $PtSe_2$ remains a semiconductor, but the energy gap decreases to 0.21 eV. Starting from a trilayer, $PtSe_2$ becomes semimetallic, as shown in Fig. 4.6. Therefore, only single-layer $PtSe_2$ is a semiconductor with a sizeable bandgap.

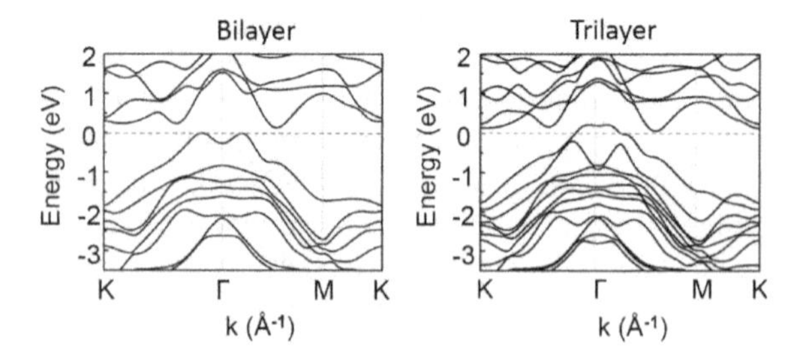

Fig. 4.6 Calculated band structures of free-standing bi- and tri-layer $PtSe_2$. Reproduced with permission from Ref. [10], © 2015 ACS

4.4 Photocatalytic Properties

The opening of a sizable bandgap within the range of visible light makes monolayer $PtSe_2$ potentially suitable for optoelectronics and photocatalysis. We explored the photocatalytic properties of monolayer $PtSe_2$ by the degradation of methylene blue (MB) aqueous solution, which serves as a typical indicator of photocatalytic reactivity [14, 15].

4.4.1 Experimental Setups

Figure 4.7a displays the schematic diagram of photocatalytic experiments. The photocatalytic activity of as-prepared monolayer $PtSe_2$ films was tested by catalytic degradation of methylene blue serving as a standard model dye under visible-light irradiation at room temperature. The ultraviolet/visible-light source was a 150 W Xe lamp located at a distance of 15 cm above the solution. A set of appropriate cut-off filters was applied to determine illumination wavelength and ensured that the photocatalytic reaction took place just under visible light. The as-prepared monolayer $PtSe_2$ films were exfoliated from the Pt(111) substrate by ultra-sonication in aqueous solution. The Pt(111) crystal was picked out from the solution to avoid its possible influence on the photocatalytic activity. Then MB molecules and ethanol (0.01 mL)

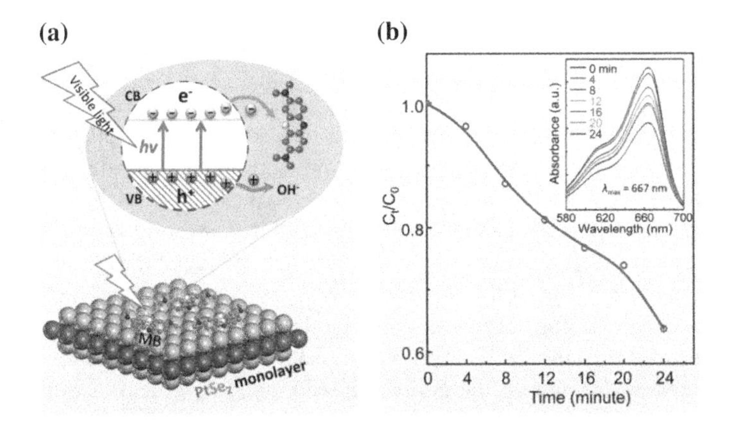

Fig. 4.7 Photocatalytic activity of a single-layer $PtSe_2$ film. **a** Schematic diagram of the photocatalytic degradation of methylene blue (MB) molecules. Electrons and holes are excited by visible-light irradiation of epitaxial $PtSe_2$ monolayer films. The MB degradation by photo-induced electrons demonstrates the photocatalytic activity of $PtSe_2$ monolayer films. **b** Time trace of the normalized concentration (C_t/C_0, where C_t and C_0 are the MB concentrations at time t min and 0 min, respectively) of the absorbance at a wavelength of 667 nm, the main absorbance peak of MB. The inset shows the UV-vis absorption spectra of MB, recorded at time intervals of 4 min. Reproduced with permission from Ref. [10], © 2015 ACS

were added to the solution containing peeled $PtSe_2$ monolayers. Before the photocatalytic reaction, the solution was kept in darkness for one hour in order to achieve an adsorption-desorption equilibrium between the $PtSe_2$ film and MB molecules. After that, the suspension was exposed to visible-light irradiation. At time intervals of 4 min, solution samples were collected and their absorbance was measured by a commercial ultraviolet-visible (UV-vis) spectrophotometer. Accordingly, the intensity changes of characteristic absorbance peaks of the MB molecules were recorded. The photocatalytic performance of $PtSe_2$ monolayers was thus evaluated by the time-dependent degradation rate C_t/C_0, as exhibited in Fig. 4.7b.

4.4.2 Photocatalytic Characterizations

The MB molecules adsorbed on the $PtSe_2$ films are degraded by electrons that are excited by visible light. The time-dependent degradation of MB with a single layer $PtSe_2$ catalyst was monitored by checking the decrease in the intensities of characteristic absorbance peaks of the MB molecules. As we can see in Fig. 4.7b, the photodegradation portion of MB molecules reached 38% after visible-light irradiation for 24 min. This rate is about four times faster than the rate obtained using $PtSe_2$ nanocrystals [16, 17], putting single-layer $PtSe_2$ in the same class as nitrogen-doped TiO_2 nanoparticles for photocatalysis [18, 19].

4.5 Valleytronics

With the existence of an energy gap in the single-layer $PtSe_2$, optical excitations can occur between the Γ point and the valley point along the Γ-M direction. This is reminiscent of the recently discovered valley-selective circular dichroism in MoS_2 by circularly polarized light [20–22]. To explore this possibility, we calculated the degree of circular polarization of free-standing single-layer $PtSe_2$. The calculated circular polarization due to the direct interband transition between the top of the valence band (VB) to the bottom of the conduction band (CB) is shown in Fig. 4.8a. It shows significant circular dichroism polarization along the M-K direction and near the Γ point. In view of the indirect energy gap, this process can be assisted by lattice vibrations. It is noteworthy that the circular dichroism polarization not only exists in the transition between the top of the VB and the bottom of the CB, it also exists in transitions to higher energy levels. Circular dichroism polarization due to transitions from the vicinity of the VB to the vicinity of the CB can be clearly seen in Figs. 4.8b–d. Due to this significant circular polarization, in the presence of a nonvanishing in-plane electric field, the anomalous charge current driven by the Berry curvature would flow to the opposite edges, leading to a valley polarized current and the resulting quantum valley Hall effect [23, 24].

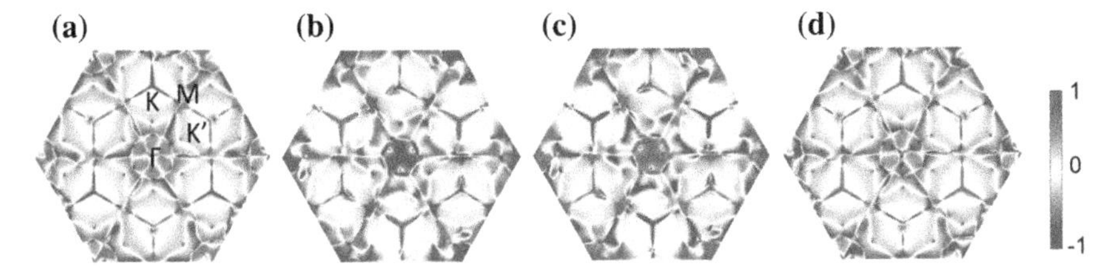

Fig. 4.8 Momentum dependence of circular polarization of single-layer $PtSe_2$. **a–d** represent the calculated circular polarization due to the direct interband transition from the top of the valence band (VB1) to the bottom of the conduction band (CB1), from VB1 to the higher conduction band (CB2), from the lower valence band (VB2) to CB1, and from VB2 to CB2, respectively. Significant circular dichroism polarization exists along the M-K direction and near the Γ point. Reproduced with permission from Ref. [10], © 2015 ACS

4.6 Summary and Outlook

We have successfully fabricated high-quality, single-crystalline, monolayer $PtSe_2$ films, a new member of the TMDs family, through a single-step, direct selenization of a Pt(111) substrate at a relatively low temperature (~270 °C).

(1) Characterizations by LEED, STM, STEM, and DFT calculations elucidated both in-plane and vertical monolayer structures with atomic resolution.
(2) The ARPES measurements and their agreement with calculations revealed the semiconducting electronic structure of the single-layer $PtSe_2$.
(3) Together with the photocatalytic performance observed experimentally, monolayer $PtSe_2$ shows promise for potential applications in optoelectronics and photocatalysis.
(4) The circular polarization of monolayer $PtSe_2$ in momentum space indicates a promising potential for valleytronic devices.
(5) Our studies are a significant step forward in expanding the family of single-layer semiconducting TMDs and exploring the application potentials of ultra-thin TMDs in photoelectronic and energy-harvesting devices.

References

1. Splendiani A et al (2010) Emerging photoluminescence in monolayer MoS_2. Nano Lett 10:1271–1275. https://doi.org/10.1021/nl903868w
2. Eda G et al (2011) Photoluminescence from chemically exfoliated MoS_2. Nano Lett 11:5111–5116. https://doi.org/10.1021/nl201874w
3. Tongay S et al (2012) Thermally driven crossover from indirect toward direct bandgap in 2D semiconductors: $MoSe_2$ versus MoS_2. Nano Lett 12:5576–5580. https://doi.org/10.1021/nl302584w
4. Zhang Y et al (2014) Direct observation of the transition from indirect to direct bandgap in atomically thin epitaxial $MoSe_2$. Nat Nanotechnol 9:111–115. https://doi.org/10.1038/Nnano.2013.277

5. Ataca C, Sahin H, Ciraci S (2012) Stable, single-layer MX_2 transition-metal oxides and dichalcogenides in a honeycomb-like structure. J Phys Chem C 116:8983–8999. https://doi.org/10.1021/jp212558p
6. Chhowalla M et al (2013) The chemistry of two-dimensional layered transition metal dichalcogenide nanosheets. Nat Chem 5:263–275. https://doi.org/10.1038/nchem.1589
7. Sipos B et al (2008) From Mott state to superconductivity in 1T-$TaS_{(2)}$. Nat Mater 7:960–965. https://doi.org/10.1038/nmat2318
8. Yang JJ et al (2012) Charge-orbital density wave and superconductivity in the strong spin-orbit coupled $IrTe_2$:Pd. Phys Rev Lett 108:116402. https://doi.org/10.1103/PhysRevLett.108.116402
9. Tongay S et al (2014) Monolayer behaviour in bulk ReS_2 due to electronic and vibrational decoupling. Nat Commun 5:3252
10. Wang YL et al (2015) Monolayer $PtSe_2$, a new semiconducting transition-metal-dichalcogenide, epitaxially grown by direct selenization of Pt. Nano Lett 15:4013–4018. https://doi.org/10.1021/acs.nanolett.5b00964
11. Zhuang HLL, Hennig RG (2013) Computational search for single-layer transition-metal dichalcogenide photocatalysts. J Phys Chem C 117:20440–20445. https://doi.org/10.1021/jp405808a
12. Guo G, Liang W (1986) The electronic structures of platinum dichalcogenides: PtS_2, $PtSe_2$ and $PtTe_2$. J Phys C: Solid State Phys 19:995
13. Dai D et al (2003) Trends in the structure and bonding in the layered platinum dioxide and dichalcogenides PtQ_2 (Q=O, S, Se, Te). J Solid State Chem 173:114–121
14. Houas A et al (2001) Photocatalytic degradation pathway of methylene blue in water. Appl Catal B 31:145–157
15. Costi R, Saunders AE, Elmalem E, Salant A, Banin U (2008) Visible light-induced charge retention and photocatalysis with hybrid CdSe-Au nanodumbbells. Nano Lett 8:637–641
16. Wilson JA, Yoffe AD (1969) The transition metal dichalcogenides discussion and interpretation of the observed optical, electrical and structural properties. Adv Phys 18:193
17. Ullah K et al (2014) Synthesis and characterization of novel $PtSe_2$/graphene nanocomposites and its visible light driven catalytic properties. J Mater Sci 49:4139–4147
18. Asahi R, Morikawa T, Ohwaki T, Aoki K, Taga Y (2001) Visible-light photocatalysis in nitrogen-doped titanium oxides. Science 293:269–271. https://doi.org/10.1126/science.1061051
19. Gole JL, Stout JD, Burda C, Lou YB, Chen XB (2004) Highly efficient formation of visible light tunable $TiO_{2-x}N_x$ photocatalysts and their transformation at the nanoscale. J Phys Chem B 108:1230–1240. https://doi.org/10.1021/jp.030843n
20. Cao T et al (2012) Valley-selective circular dichroism of monolayer molybdenum disulphide. Nat Commun 3:887
21. Mak KF, He K, Shan J, Heinz TF (2012) Control of valley polarization in monolayer MoS_2 by optical helicity. Nat Nanotechnol 7:494–498. https://doi.org/10.1038/nnano.2012.96
22. Zeng H, Dai J, Yao W, Xiao D, Cui X (2012) Valley polarization in MoS_2 monolayers by optical pumping. Nat Nanotechnol 7:490–493. https://doi.org/10.1038/nnano.2012.95
23. Rycerz A, Tworzydło J, Beenakker C (2007) Valley filter and valley valve in graphene. Nat Phys 3:172–175
24. Xiao D, Yao W, Niu Q (2007) Valley-contrasting physics in graphene: magnetic moment and topological transport. Phys Rev Lett 99:236809

Chapter 5
Conclusions and Prospect

Abstract As a summary of contents, the major research results and outlooks are included in this chapter.

Keywords Germanene · Hafnium honeycomb · $PtSe_2$ · 2D materials

Graphene was most highly studied in material-related science in the past dozen years because of its exceptional properties and wide applications. More importantly, the rise of graphene initiated a new field—two dimensional (2D) materials. The explosive studies on graphene have inspired considerable interests towards graphene-analogous 2D materials. It is in this background that this thesis aims at investigating novel 2D atomic crystals beyond graphene. A variety of surface-science techniques including LEED, STM, STEM, XPS, ARPES, and Raman spectroscopy combined with DFT calculations are employed to systematically study three new graphene-like 2D materials—germanene, hafnene, and $PtSe_2$. The major research results and outlooks are summarized as follows.

(1) The experimental synthesis of germanene, a germanium-based counterpart of graphene, is reported for the first time, which is the third single-element 2D material following graphene and silicene. A ($\sqrt{19} \times \sqrt{19}$) superstructure with respect to the Pt(111) substrate is observed by LEED and STM, which coincides with the (3×3) superlattice of a buckled germanene sheet demonstrated by the first-principle calculations. The calculated electron localization function reveals that adjacent Ge atoms directly bind to each other, confirming the continuity of 2D germanene sheets on the Pt(111) surface. This work develops the first method of preparing single-layer germanene and facilitates the subsequent experimental studies on its physical properties and potential applications.
(2) The first honeycomb structure composed of transition metals-hafnene-is experimentally fabricated on an Ir(111) surface using UHV-MBE method. Theoretical calculations reveal interesting ferromagnetism, high density of d states at the Fermi level, and large spin-splitting of the Dirac cones for a freestanding Hf honeycomb structure, indicating rich electronic, magnetic and catalytic potential

© Springer Nature Singapore Pte Ltd. 2020

L. Li, *Fabrication and Physical Properties of Novel Two-dimensional Crystal Materials Beyond Graphene: Germanene, Hafnene and* $PtSe_2$, Springer Theses,
https://doi.org/10.1007/978-981-15-1963-5_5

applications. The realization and investigation of hafnene broaden the scope of artificially synthesized honeycomb structures and offer a new platform for studying hitherto unknown quantum phenomena and electronic behaviors in 2D systems.

(3) Single-layer $PtSe_2$ films, a new transition metal dichalcogenide (TMD) crystal, are prepared through direct selenization of a Pt(111) substrate. Multiple experimental characterizations and DFT calculations depict the structural and electronic properties of monolayer $PtSe_2$. The high photocatalytic activities measured experimentally make $PtSe_2$ single layers promising materials in optoelectronics and photocatalysis. Moreover, monolayer $PtSe_2$ exhibits circular polarization in momentum space, indicating potential applications for valleytronic devices. These studies are a significant step forward in enriching the family of single-layer semiconducting TMDs and exploring their application potentials in photoelectronic and energy-harvesting devices.

CPSIA information can be obtained
at www.ICGtesting.com
Printed in the USA
LVHW081451030220
645675LV00002B/7

9 789811 519628